Lead-Based Paint
Handbook

TOPICS IN APPLIED CHEMISTRY

Series Editors: Alan R. Katritzky, FRS
Kenan Professor of Chemistry
University of Florida, Gainesville, Florida

Gebran J. Sabongi
Laboratory Manager, Encapsulation Technology Center
3M, St. Paul, Minnesota

BIOCATALYSTS FOR INDUSTRY
Edited by Jonathan S. Dordick

CHEMICAL TRIGGERING
Reactions of Potential Utility in Industrial Processes
Gebran J. Sabongi

THE CHEMISTRY AND APPLICATION OF DYES
Edited by David R. Waring and Geoffrey Hallas

HIGH-TECHNOLOGY APPLICATIONS OF ORGANIC COLORANTS
Peter Gregory

INFRARED ABSORBING DYES
Edited by Masaru Matsuoka

LEAD-BASED PAINT HANDBOOK
Jan W. Gooch

STRUCTURAL ADHESIVES
Edited by S. R. Hartshorn

TARGET SITES FOR HERBICIDE ACTION
Edited by Ralph C. Kirkwood

A Continuation Order Plan is available for this series. A continuation order will bring delivery of each new volume immediately upon publication. Volumes are billed only upon actual shipment. For further information please contact the publisher.

Lead-Based Paint Handbook

Jan W. Gooch

Georgia Institute of Technology
Atlanta, Georgia

Plenum Press • New York and London

05592662
CHEMISTRY

Library of Congress Cataloging-in-Publication Data

Gooch, Jan W. (Jan Woodall), 1946-
    Lead-based paint handbook / Jan W. Gooch.
        p.   cm. -- (Topics in applied chemistry)
    Includes bibliographical references and index.
    ISBN 0-306-44448-8
    1. Lead abatement. 2. Lead based paint.   I. Title. II. Series.
TD196.L4G66   1993
628.5'2--dc20   93-1831
                                                                    CIP

ISBN 0-306-44448-8

©1993 Plenum Press, New York
A Division of Plenum Publishing Corporation
233 Spring Street, New York, N.Y. 10013

I wish to acknowledge the following people
for their contributions to this book:
Garth B. Freeman, John T. Sparrow, Paul M. Hawley,
Christopher D. Papanicolopulos, and
Cerena L. Gooch

Preface

Lead-based paint has become a national issue and will continue to be a high-priority focus of national, state, and local agencies until there is no lead-based paint in the United States. Lead-based paint has become a tremendous health hazard for people and animals. Lead-based paint has been in widespread use throughout Europe and the United States. Lead has been known to be a health hazard since the time of Pliny the Elder (A.D. 23–79), but it was deemed that the advantages of lead in paint outweighed the health hazards. There has been a change in outlook, and in 1973 the U.S. Congress banned all lead paint from residential structures. A voluminous number of law suits have been initiated since, and continue to be litigated with the purpose of determining the parties responsible for the lead poisoning of children and others and to exact the indemnities.

Lead-based paint is still authorized for use on bridges and nonresidential structures, and thousands of city, state, military, and federal government housing projects still contain lead-based paint. This paint must be removed if these dwellings are to be safe living quarters, especially for children. Abatement techniques continue to be evaluated; some have been used successfully. Lead-based paint abatement will continue into the next century, and it is hoped that this comprehensive volume will serve as a guide for those seriously interested in this important subject.

Organizational Guide

This first comprehensive handbook on lead-based paint is carefully structured to give the engineer, scientist, or attorney an understanding and a guide for detecting, analyzing, and abating lead-based paint. The handbook is generally organized into five major subject areas: (1) regulations for lead-based paint, (2) paint history and related materials, (3) lead and toxicology, (4) analysis of paint and related components, and (5) abatement of lead-based paint.

The organization of this handbook is outlined in the table of contents which will assist the reader in quickly locating the topic of interest within the text. Each major topic is thoroughly discussed with its relevance to lead-based paint together with references. Extensive historical backgrounds are included on paint and its constituents (lead, toxicology of lead, and the manufacture of white lead pigments in the United States). The historical information is important, since most of the lead-based paint in the United States was used during the period 1900–1973. The manufacture of lead-based paint for residential purposes was ended in 1973 by an act of Congress.

The reader can conveniently find a method of analysis for pigments and binders in the table of contents and locate the method within the text. Each method is explained and discussed sufficiently to familiarize the reader with its function and purpose. Extensive attention is given to identifying those paint binders and pigments known to be widely used in lead-based paint, including appended infrared spectra and X-ray powder files. This information is taken from the author's experience in abatement and legal investigations.

Finally, a comprehensive discussion on lead-based paint abatement is included together with studies, methods, and tools. The bulk of the reported information was generated from abatement research and actual housing projects in the United States.

Contents

4. Lead Pigments

5. Health Problems Associated with Lead-Based Paint

6. Field Detection Methods

7. Laboratory Analysis Methods

8. Methods of Analyzing Vehicles

9. Analyzing Dust and Soil for Lead

10. General Analysis of Lead-Based Paint

11. Regulation and Abatement of Lead-Based Paint

12. Disposing of Lead Paint Waste

1

Introduction to Lead-Based Paint

1.1. SOURCES OF LEAD-BASED PAINT AND REGULATIONS

1.1.1. Sources of Lead-Based Paint

Residential houses, commercial and industrial buildings, and bridges constitute the majority of lead-based painted surfaces. Before the 1970s, the interior and exterior surfaces of houses and buildings were painted with lead-pigmented paint. Bridges continue to be painted with lead based paint. There have been specific advantages to using lead-based paint: It is an excellent natural mildewcide, corrosion inhibitor, and color stabilizer.* The typical sources of lead from painted surfaces are chipped and peeled particles. Particle sizes range from 1 mil to several millimeters. They do not become airborne particles unless they are pulverized by continuous pedestrian traffic or other mechanical methods of shearing the particles. Lead-based paint is usually old and brittle and therefore susceptible to shearing and separating into smaller particles and eventually into individual pigment particles (nominally $0.1–25$ μ in diameter, as measured by the author).

1.1.2. Review of Lead-Based Paint Regulations

New data (CDC, 1991) indicate significant adverse effects of lead exposure in children at blood lead levels previously believed to be safe. Some adverse health effects have been documented at blood lead levels at least as low as 10 μg

---

*Coastal areas in the United States are particularly vulnerable to deterioration by mildew, and lead-based paint is unsurpassed for thwarting propagation of mildew.

per deciliter of whole blood. The 1985 intervention level of 25 μg/dl is therefore being revised downward to 10 μg/dl, and a multitier approach has been adopted.

Primary prevention efforts (that is, eliminating lead hazards before children are poisoned) must receive more emphasis as blood lead levels of concern are lowered. The goal of all lead-poisoning prevention activities should be to reduce children's blood lead levels below 10 μg/dl. If many children in the community have blood lead levels ≥ 10 μg/dl, communitywide intervention (primary prevention activities) should be considered by appropriate agencies. Intervention for individual children should begin at blood lead levels of 15 μg/dl.

Childhood lead poisoning is one of the most common pediatric health problems in the United States Today. Lead poisoning, for the most part, is silent: most poisoned children have no symptoms; the vast majority of cases therefore go undiagnosed and untreated. Lead poisoning is widespread: It is not solely a problem of inner-city or minority children. No socioeconomic group, geographic area, nor ethnic population is spared.

"Preventing Lead Poisoning in Young Children" (Roper, 1991) provides guidelines on childhood lead poisoning prevention for diverse groups: public health programs screening children for lead poisoning; pediatricians and other health care practitioners; and government agencies, elected officials, and private citizens seeking what constitutes a harmful level of lead in blood. However, it is not possible to select a single number to define lead poisoning for the various purposes of all of these groups. Epidemiologic studies have identified harmful effects of lead in children at blood lead levels at least as low as 10 μg/dl. Some studies have suggested harmful effects at even lower levels, but the body of information accumulated so far is not adequate for effects below about 10 μg/dl to be evaluated definitively. As yet, no threshold has been identified for the harmful effects of lead.

Because 10 μg/dl is the lower level of the range at which effects are now identified, primary prevention activities—communitywide environmental interventions and nutritional and educational campaigns—should be directed at reducing children's blood lead to at least below 10 μg/dl. Blood lead levels between 10–14 μg/dl are in a border zone. While the overall goal is to reduce children's blood lead levels below 10 μg/dl, there are several reasons for not attempting medical interventions with individual children to lower blood lead levels of 10–14 μg/dl. (1) At low blood lead levels, laboratory measurements may be inaccurate or imprecise, so a blood lead level in this range could, in fact, be below 10 μg/dl. (2) Effective environmental and medical interventions for children with blood lead levels in this range have not yet been identified and

Table 1.1.   Interpretation of Blood Lead Test Results and Follow-up Activities:
Class of Child Based on Blood Lead Concentration

Class	Blood Lead Concentration (μg/dl)	Comment
I	≤9	A child in Class I is not considered to be lead poisoned.
IIA	10–14	Many children (or a large proportion of children) with blood lead levels in this range should trigger communitywide childhood lead-poisoning prevention activities. Children in this range may need to be rescreened more frequently.
IIB	15–19	A child in Class IIB should receive nutritional and educational interventions and more frequent screening. If the blood lead level persists in this range, environmental investigation and intervention are necessary.
III	20–44	A child in Class III requires environmental evaluation, remediation and a medical evaluation. Such a child may need pharmacologic treatment for lead poisoning.
IV	45–69	A child in Class IV requires both medical and envronmental intervention, including chelation therapy.
V	≥70	A child exhibiting Class V lead poisoning constitutes a medical emergency. Medical and environmental management must begin immediately.

Source: CDC (1991).

evaluated. (3) The sheer number of children in this range would preclude effective case management and detract from the individualized follow-up required by children who have higher blood lead levels.

The single, all-purpose definition of childhood lead poisoning has been replaced with a multitier approach, described in Table 1.1. Guidelines recommend community prevention activities triggered by blood lead levels ≥ 10 μg/dl and evaluation, environmental investigation, and remediation for all children with blood lead levels ≥ 20 μg/dl. All children with blood lead levels ≥ 15 μg/dl should receive individual case management, including nutritional and educational interventions and more frequent screening. Furthermore, depending on the availability of resources, environmental investigation (including a home inspection) and rcmediation should be done for children with blood lead levels of 15–19 μg/dl if such levels persist. The highest priority should continue to be children with the highest blood lead levels.

Other differences between the 1985 and 1991 statements include the following points:

- Test screening choice. Because the erythrocyte protoporphrin levels is not sensitive enough to identify children with elevated blood lead levels below about 25 μg/dl, the preferred screening test is now blood lead measurement.
- Universal screening. Since virtually all children are at risk for lead poisoning, universal screening is recommended except in communities where a large percentage of children have been screened and found not to have lead poisoning. Full implementation of a universal screening program requires the ability to measure blood lead levels on capillary samples and the availability of cheaper and easier to use methods of blood lead measurement.
- Primary prevention. Focus efforts on preventing lead poisoning before it occurs. This requires communitywide environmental interventions, as well as educational and nutritional campaigns.
- Succimer. In January 1991, the U.S. Food and Drug Administration approved succimer, an oral chelating agent, for chelating children with blood lead levels over 45 μg/dl.

Childhood lead-poisoning prevention programs have had a tremendous impact on reducing the occurrence of lead poisoning in the United States. Because of these programs, deaths from lead poisoning and lead encephalopathy are now rare. These programs have targeted high-risk children for periodic screening; provided education to caretakers about the causes, effects, symptoms, and treatments of lead poisoning; and ensured medical treatment and environmental remediation for poisoned children. Screening and medical treatment for poisoned children will remain critically important until environmental sources most likely to poison children are eliminated.

Federal regulation and similar actions have resulted in substantial progress in reducing blood lead levels in the entire U.S. population. For example, in the last two decades, the virtual elimination of lead from gasoline has resulted in reduced blood lead levels in children and adults. Lead levels in food have also decreased since most manufacturers stopped using leaded solder in cans and since atmospheric deposition of lead on food crops declined with the reductions of lead in gasoline. In 1978, the Consumer Product Safety Commission banned the addition of lead to new residential paint.

Important environmental sources and pathways of lead still remain, with lead-based paint and lead-contaminated dusts and soils as the primary sources and pathways of lead exposure for children. In addition, children continue to be exposed to lead through air, water, and food, as well as occupations and hobbies of parents and caretakers. The focus of prevention efforts must therefore expand from merely identifying and treating individual children to include primary prevention—preventing exposure to lead before children become poisoned. This requires a shared responsibility by many public and private agencies. Public agencies will have to work with pediatric health care providers to identify communities with childhood lead-poisoning prevention problems and unusual sources of lead and to ensure environmental follow-up of poisoned children. Public housing and economic development agencies will have to integrate lead paint abatement into housing rehabilitation policies and programs. Health care providers will have to phase in virtually universal screening of children. Public and private organizations must continue to develop economical and widely available blood lead tests to make such screening possible. Public and private housing owners must bear a portion of the financial burden for abatement.

Efforts have begun to increase federal support of childhood lead-poisoning prevention programs and follow-up activities. Ongoing efforts to develop infrastructure and technology by the public and private sectors include (1) inexpensive, ease-to-use portable methods of measuring blood lead levels, (2) training and certification programs for lead paint inspectors and abatement contractors, and (3) developing and testing new abatement methods, including encapsulants. Changes in the approach to lead poisoning are not meant to increase the emphasis on screening of children; the long-term goal is still prevention. Until primary prevention of childhood lead poisoning can be achieved, however, increased screening and follow-up of poisoned children are essential.

1.1.3. Eliminating Childhood Lead Poisoning

In February 1991, the U.S. Department of Health and Human Services released a Strategic Plan for the Elimination of Childhood Lead Poisoning (HHS, 1991). This plan describes the first 5 years of a 20-year societywide effort to eliminate lead poisoning. It places highest priority on first addressing the children at greatest risk for lead poisoning. The U.S. Department of Housing and Urban Development (HUD, 1990) and the Environmental Protection Agency (EPA, 1991) have both released plans dealing with the elimination of lead hazards. Eradicating this disease will require a tremendous effort from all levels of government as well as the private sector.

Removing lead paint was mandated by different federal regulations, starting in 1973 when the Consumer Product Safety Commission (CPSC) established the maximum lead content in paint to be 0.5% by weight (HHS, 1991) in a dry film of paint. In 1978, the CPSC lowered the allowable lead level to 0.06% by weight. The Housing and Urban Development (HUD) guidelines require paint to be removed when lead is 1 mg/cm^2 or 0.5% by weight (HUD, 1990) (which ever is more stringent). Certain states and local (county) authorities have adopted different action levels—1.2 or 0.7 mg/cm^2. The action level of lead-in-soil has been unofficially adopted by the Environmental Protection Agency (EPA) to be approximately 500 parts per million (EPA, 1991).

1.2. SOURCES OF LEAD IN THE ENVIRONMENT

1.2.1. General Sources of Lead

Various sources of lead are listed in Table 1.2, including primary and secondary lead-smelting operations; manufacturing storage batteries and reclaiming batteries; emissions from leaded gasoline; manufacturing iron and steel; municipal solid waste incinerators; coal and oil combustion; lead solder and lead water pipes; producing ammunition, lead sinkers, and other products containing lead; and using lead as pigments in paint and printed materials. Although the use of lead in gasoline, solder, water pipes, and residential paint has been substantially reduced, lead from these uses remains in the environment.

Depositions from Civil Action No. 87-2799-T in U.S. District Court for the District of Massachusetts (1991) contain current lead levels from various sources in Boston residential areas. Lead in soil was highlighted at the 1986 Trace Substances in Environmental Health Conference held in Columbia, Missouri. As a result, a special conference on "Lead in Soil: Issues and Guidelines" was held in 1988 under the sponsorship of various scientific, regulatory, industrial, and educational associations. The Society for Environ-

Table 1.2. Sources of Lead in the Environment

Nonairborne	Airborne
Lead-based paint in buildings, interiors and exteriors	Oil, coal, combustion and solid waste
Lead food cans	Metal refining, smelting, and manufacturing processes emissions
Lead gasoline and dust	Emissions from alkyl lead manufacturers
Lead water pipes	Lead battery manufacturers and recyclers
Soil	Portland cement manufacturing
	Painted surfaces weathering

mental Geochemistry and Health (SEGH) was requested to form a special task force to review the most current scientific literature on human lead exposure, and in particular, the role of lead in soil, and to recommend appropriate guidelines. The goal of the SEGH task force guidelines was to provide a methodology by which public health officials, regulatory agencies, industrial environmental managers, and others concerned with lead exposure can assess potential human exposure associated with various soil lead levels.

The Twenty-Fifth Annual Conference on Trace Substances in Environmental Health was held at the University of Missouri, May 20–23, 1991. Findings and recommendations conclude that (1) lead is ubiquitous in the human environment; (2) any lead exposure assessment must analyze all potential sources and exposure pathways; and (3) acceptable soil lead levels can be determined only through an approach that takes into account numerous factors including but not limited to the population at risk, the physical availability of the lead, and the bioavailability of the lead, which is the amount of lead that can be ingested into the blood stream.

Lead has had and continues to have many beneficial uses. Given its past and present widespread industrial use, there are numerous sources that introduce lead into the environment, thus making it accessible to human beings. Human beings are exposed to lead in the air they breathe, the water they drink, and the food they eat. Current literature also identifies the ingestion of surface dust and soil as an important pathway of lead exposure for young children (ATSDR, 1988). Ingesting lead-based paint may also be a source of childhood exposure.

Each of these potential pathways has multiple sources of lead contamination. The actual lead exposure experienced by any individual varies greatly depending on many specific factors. Because there are multiple sources and pathways of lead exposure, no single source accounts for the occurrence of elevated blood lead levels in children.

The total amount of lead is an accumulation from these combined sources. In addition, airborne lead may eventually settle on accessible soil and/or dust and thus become part of an additional exposure pathway. Even after airborne lead settles on streets, sidewalks, and nearby soil, it can become airborne again if it is disturbed by, for example, vehicular or pedestrian traffic or excavation activities. The relative importance of any of these airborne lead sources depends on site-specific exposure factors. Research has shown a relatively clear correlation between airborne lead concentrations and blood lead levels (ATSDR, 1988).

There have been numerous efforts to reduce human, and especially childhood, lead exposure. Between 1976 and 1980, the national mean blood lead levels declined by 37% from 15.9 µg/dl to 9.6µg/dl (Annest, 1983). This decline has been primarily attributed to the corresponding reduction of the use of lead in gasoline. The reduction of lead in food and water and the regulation of industrial emissions have also helped reduce the mean blood lead levels. While the national mean blood lead level in the mid-1970s was approximately 16 μg/dl, data indicate that mean blood levels for inner-city children during that time was much higher and in fact averaged 20–25 μg/dl (Stark, 1982). Some studies have reported that during the late 1960s and early 1970s, blood lead levels of inner-city children averaged 40–50 μg/dl (Billick, 1983). During this time, lead levels were much higher than they are today (AAP, 1969).

During the past 10–15 years, the medical and scientific communities began to focus on the ingestion of lead-contaminated soil and dust as the principal pathway of exposure for young children, given the normal hand-to-mouth activity associated with early childhood development. Children also suck on fingers, toys, and other objects that can become covered with lead-contaminated dust and soil. In fact, for the majority of children, exposure to lead by ingesting lead-contaminated soil and dust is more frequent than ingesting lead-based paint residue (Boggess and Wixon, 1977).

1.2.2. Lead in the Water Supply

The drinking water in 130 cities, including New York, Detroit, and Washington, D.C., contains excessive levels of lead according to the EPA (Gutfeld, 1992). The EPA's initial sampling of 660 large public water systems indicated that 32 million Americans drink water from systems that failed the federal test of 15 parts per billion. In 10 cities, lead was above 70 parts per billion, with Charleston, South Carolina, posting the worst level at 211 parts per billion. These findings do not represent average lead levels. Samples were taken only from high-risk homes—those served by lead service lines or that contain lead interior piping or copper piping with lead solder. The tests were also based on first-draw water before the system was flushed. The latest results were based on sampling between January and June 1992.

Lead makes its way into the water supply primarily through the use of old lead water pipes or lead solder in plumbing. Lead in drinking water can be a significant source of human exposure because of the large amounts of water consumed each day and the ready absorption of lead from water. Lead water pipes are common in older housing and the water distribution systems of older cities. The contribution of lead pipes to blood levels is well documented (ATSDR, 1988). Because soft water is likely to corrode metallic surfaces, lead water pipes are a significant source of lead exposure for people in soft water areas. It also appears that the water supply in the nation's public schools contains lead and thus provides an additional source of lead exposure.

1.2.3. Lead in Food and Containers

Human beings are also exposed to lead in food through food cans with lead-soldered seams or through lead-glazed dishes, lead cookware, and airborne lead deposits that contaminate crops. Like water, lead in food can be a significant source of exposure because of the large quantities consumed on a daily basis. Dietary lead intake varies depending on such factors as geographical location, agricultural practices, socio-economic conditions, and food preparation practices.

The U.S. Food and Drug Administration (FDA) reported that in 1979, over 90% of food containers contained lead-soldered seams. The FDA has since sought to reduce a child's daily dietary lead intake by, for example, establishing permissible lead residues in evaporated milk and evaporated skim milk and reducing lead in canned infant formulas, infant fruit, and vegetable juices. By 1986, the percentage of food containers with lead-soldered seams was reduced to 20%. (ATSDR, 1988; CDC, 1985; EPA, 1988).

1.2.4. Lead in Paint, Printed Material, and Cosmetics

Ingesting residues from lead-based paint and printed material containing lead pigments is another source of childhood exposure to lead. Lead in cosmetics and some home remedies are sources of human exposure.

1.2.5. Leaded Gasoline

Soil acts as a sink, which over time accumulates lead deposits from various sources, such as leaded gasoline emissions, emissions from metal refining and smelting operations; industrial emissions, including oil, coal, and solid waste combustion; and weathered lead-based paint. Therefore even though lead has been eliminated as a gasoline additive in the United States, lead from the combustion of leaded gasoline is still present at significant levels in urban soil and dust.

The significant contribution of leaded gasoline emissions and industrial sources to soil lead levels in urban areas is amply illustrated in the literature. In terms of the gross amount of lead that enters the environment, nearly twice as much lead was used in leaded gasoline than to produce white lead pigments (see U.S. Department of Interior Bureau of Mines Reports). Also, lead emitted into the air through the combustion of leaded gasoline consists of about $0.01-1.5$ μ particulate lead, which is disbursed in the surrounding environment and settles as street, curbside, and soil dust. Much of the lead used in paint pigment, however, remains intact and may be covered by layers of nonlead-based paint, wallpaper, or other coverings and therefore is inaccessible.

1.2.6. Lead in Soil

Mielke *et al*. (1983) found that lead as well as cadmium, copper, nickel, and zinc were concentrated and ubiquitous in garden soil in the Baltimore, Maryland inner city. The authors noted that elevated soil levels did not appear to be related to painted surfaces weathering in the inner city and concluded that emissions from vehicular traffic was likely a major source of the elevated soil lead levels. Mielke *et al*. (1989) confirmed the contribution of leaded gasoline emissions to urban soil by studying soil lead levels in four cities in Minnesota. On the basis of that research, the authors concluded:

> The major patterns of soil lead in Minnesota are related to city size and traffic flow. The accumulation of leaded petrol is directly linked to environmental contamination and lead exposure of childhood populations in Minnesota.

The Minnesota research suggested that even soil lead levels near the foundations of inner-city buildings were the result of airborne lead particles that had become impacted on the sides of buildings and subsequently washed off by precipitation. Again, this research revealed that the presence of lead-based paint was not necessarily the primary source of elevated soil lead levels in the inner-city areas (Fergurson, 1986).

1.2.7. Lead in Ambient Air

The study of ambient air lead concentrations has also shown the significant correlation between the combustion of leaded gasoline and soil lead concentrations. The EPA has noted that air lead concentration decreases as you travel from the center of a city and that soil lead levels are a direct function of the deposition of ambient air lead (see EPA reports for 1986). The contribution of leaded gasoline emissions and soil lead levels to interior house dust is evident in studies showing that lead in house dust is related to exterior lead levels (Chaney, 1989).

One of the largest sources of lead in ambient air since the 1920s in the United States has been from the combustion of leaded gasoline. However, the combustion of used motor oil has been reported as the nation's largest single source of airborne lead by the EPA Hazardous Waste Treatment Council, November 1991.* Used motor oil is collected nationwide at service stations and burned in boilers (except in California) to generate heat. The motor-oil-recycling industry is regulated by the National Oil Recyclers Association in Cleveland, Ohio.

*See also *USA Today*, Nov. 13, 1991, p. 10A.

1.2.8. Bioavailability

One of the important factors in determining the contribution of lead in soil and dust to lead levels in children is bioavailability (Chaney, 1989). Factors influencing how much lead in soil and dust ingested by children may be absorbed into the blood stream include the physical and chemical properties of the lead, particle size, and the nutritional status of the particular child. Lead from leaded gasoline emissions are predominately small particles and therefore readily bioavailable to children (Chaney, 1989). The bioavailability of lead in paint varies with particle size and the type of lead compound present in the paint (*e.g.*, lead carbonate, lead sulfate).

In one case history, Boston recognized the contribution of elevated soil lead levels in the city to childhood blood lead levels when it sought EPA assistance in removing soil from certain areas determined to be lead-in-soil hot spots. In 1985, the Boston Department of Health and Hospitals Office of Environmental Affairs submitted a request to the EPA entitled, "Boston Child Lead Poisoning: Request for Immediate Cleanup of Lead-Contaminated Soil in Emergency Areas." In this report, Boston identified 28 residential areas in Dorchester, Mattapan, Roxbury, and Jamaica Plain, each averaging two to three blocks, with disproportionately high numbers of children with elevated blood lead levels associated with lead-in-soil hot spots. The report states that while the average soil lead level in Boston is 600–700 ppm, soil samples collected and analyzed by the city revealed that the average soil lead level at the homes of the children with elevated blood lead levels was approximately 2000 ppm. The Boston lead-in-soil report identifies a significant source of lead exposure for these children.

While lead in soil does not appreciably dissipate over time, some lead in soil and especially lead in street and sidewalk dust can be washed away by rain. It is likely that the lead concentrations in street and curbside dust were greater in the early 1970s during the peak use of leaded gasoline than it is today (Rolfe *et al.*, 1975; Wheeler *et al.*, 1979; Hamilton *et al.*, 1984; and Brunekreef *et al.*, 1983).

2

History of Paint and Coatings Materials

2.1. EARLIEST PAINTS

2.1.1. 20,000 B.C.

You must know the history of paint and related materials to be able to distinguish between paint prepared 1 year or 25 millennia ago. The earliest reported paints are from Europe and Australia (Boatwright, 1990). European paints were produced by Neanderthal and/or Cro-Magnon man; Australian paints were produced by the same Aborigines that inhabit Australia today. Both types of paints date back to approximately 20,000 years B.C. While many of the drawings are monochromatic, others represent an ingenious palette of colors made from natural earth pigments, many of which we still use, such as iron oxide, chalk, charcoal, and terra verde. Many of the paints were applied with the fingertips, while others were applied with primitive brushes made by chewing on the tips of soft twigs until the ends frayed, forming rudimentary bristles. Rudimentary binders included animal fats, blood, egg whites and yolks.

2.1.2. 9000 B.C.

On the North American continent (Boatwright, 1990), a race of people living west of the Pecos River in Texas about 9000 B.C. also learned how to make a primitive paint. These people used their primitive paints in much the same manner as their European and Australian counterparts to paint the rock walls of their living quarters with pictures of animals and people. Their paints were also made of natural pigments and whatever binders they could procure.

13

2.1.3. Biblical Times

Turning to another civilization and the history of paint, we find that Noah was admonished to "pitch the Ark within and without with pitch" (Gen. 6:14). The word pitch refers to a liquid asphalt that seeped out of the ground; today, on the island of Trinidad, there are still areas where asphalt seeps out of the ground. We may not normally think of asphalt as a paint, but we must remember that it is a coating, which makes it relevant to the history of paint.

Several thousand years after Noah, the Egyptians began to make and use paint. They too used many natural pigments and materials in their paints, and they were the first to develop synthetic pigments. What was known as Egyptian blue was a composition of lime, alumina, silica, soda ash, and copper oxides. It was made, according to Vitrivius, by calcining a mixture of sand, soda, and copper. Their natural colors were derived from red and yellow ochres, cinnabar, hematite, orpiment, burnt yellow marl, gold leaf, malachite, azurite, charcoal, lampblack, and gypsum. The Egyptians also developed an organic pigment that involved a madder carmine on a gypsum base. They employed a variety of organic and inorganic materials as their binders: gum arabic, egg whites and yolks, gelatin, and beeswax. Lime plaster and plaster of Paris were also used as binders. Their ships were coated with both asphalt and balsam exudations.

2.1.4. Ancient Egypt

The Egyptians progressed further in manufacturing and using paints than did any of their predecessors. They decorated their coffins with lime paints, and at one time used a coating of varnish similar to Venice turpentine of Canada balsam as the final coating. Temple stones were highly colored by rubbing pigments into the stones just as stoves were blackened in the earlier part of this century.

2.1.5. Ancient Orient

The Orientals developed their own techniques as time passed. The art of suspending pigments in water, with or without a binder, was common. The Persians used gum arabic as a binder, while the Chinese used a glue. In India, crayons were made from boiled rice, and colors were applied with brushes and a crude stylus. East Indians used cinnabar, lac dye, red ochre, lampblack, and lime for pigments. Lime was blended with the preceding colors to produce pastels of one kind or another. The Chinese and Japanese used cinnabar, azurite, copper carbonate, malachite, ultramarine, indigo lake, orpiment, lime, red lead, litharge, lampblack, black ochre, gold powder, and brown coral powder as colors. Gold powder was used only in drawings of Buddha, since yellow was the sacred color of the Orient.

The natural fungicide hinolitol is better known in the United States as thujaplicin. Thujaplicin belongs to the family of tropolones, which are rings analogous to benzene, except that the ring contains seven carbons. Thujaplicin and its related compound, thujic acid, are currently being tested in the United States by the Chicago Society for Coatings Technology (Boatwright, 1990). These compounds have been, and may still be, in the process of being tested in Jamaica by some paint firms. The hinoki tree, which contains a natural fungicide, has been used by the Japanese in temples for thousands of years with no treatment, coating of any kind, or any other assistance.

2.1.6. Prehistoric American Indian

The American Indian also developed paint but exactly when is unknown; possibly it was at the same time as the Greeks and Romans. Indians, like other early people, used mixtures of various natural oxides, charcoal, chalk, *etc.*, to make paints. Indians also used animal fats, blood, egg white and yolks, as a binder. Although the Indians did not decorate caves (at least none have been found), the braves did decorate their faces with war paint, the exterior of the tepees were decorated. When the first settlers landed, they found the Indians still using the same paints.

2.1.7. Ancient Greeks and Romans

From the birth of Christ on, early artists were adept at making and using paints, particularly the Greeks and Romans. Early Cretan and Etruscan (Dioscorides, A.D. 40–90) were done in fresco, with glue and albumin as binders. While frescos abound, no Roman or Greek paints on panels or canvas have been found; Pliny the Elder describes the materials and techniques used in painting. Besides the pigments common to the Egyptians, the Romans made several artificial colors from white lead, litharge, red lead, yellow lead oxide, verdigris, and bone black. Carmine pigments were made from woods, kermes, and madder precipitated on a white earth, or mixed with honey and chalk for painting. Pitch was used during this period to seal ship timbers, and a mixture of wax and pitch was used for ship bottoms. Resins and oils were used only for liniments, and there is no record of any varnish being made, except for the bituminous coatings made with pitch.

2.1.8. First Shellac

Shellac, which was erroneously called Indian amber by Pliny (A.D. 23–79), is made from one of the few resins obtained from insects. This resin, called lac, is secreted by a coccid insect (*Laccifer lacca*) that feeds on the lac trees in India and Thailand. The resin was used in making lac sticks to coat rotating objects on a lathe over 3000 years ago.

Lac, the hard resinous secretion, is dissolved in ethanol, and residue from insects and twigs is removed by filtration. The color, which ranges from light yellow to dark orange, depends on the type of host tree used as a source of the shellac and the extent of refining. Since shellac is insoluble in aliphatic hydrocarbon solvents, such as mineral spirits, it continues to be used as a stain suppressant seeker under other solvent-based coatings. Shellac is also used to coat pills, candy, and fruit.

2.1.9. First Lacquer

Concurrently, the Egyptians, Japanese, and Chinese were beginning to develop lacquers (Stillman, 1960).* Some time before 200 B.C., the Chinese used the exudation (sap) from the conifer *Rhus vernicifera* (which became known as the sumac or varnish tree) as a coating. This plant has also been called the urushi tree. The tree belongs to the same family as the poison ivy plant, and like it, all parts of the tree are toxic—tree, sap, and latex. Those who tap the tree must wear gloves and protective clothing. The active irritant is urushiol, a catechol derivative.

This latex is not really the sap of the tree but a healing compound from wounds to the tree. Contrary to most paints and coatings, the lacquer made from this latex requires very high humidity for curing to take place. Lacquers from this latex have produced coatings that are expected to last thousands of years.

The first true lacquer was developed in Japan from the sap of a sumac tree (*Rhus vernicera*) during the Chou dynasty about 3000 years ago. Japanese lacquers are a type of oleoresin, which dries by oxidation in a damp atmosphere.

The use of Japan lacquers was expanded and improved during the Ming dynasty (1362–1644). The resin, obtained by thermal evaporation of the aqueous solution, was mixed with pigments and used as a high-gloss coating. It was not uncommon to build up a thick coating by applying as many as 250 layers of lacquer.

---

*The word lacquer was derived from the French word lecre for resin, which was derived from the Latin word lac for milk.

The Chinese added carbon black to the lacquer and first used it in the second century B.C. to write on bamboo strips. They also applied it as a paint for pottery and by the second century A.D., to buildings and musical instruments. The Chinese obtained carbon black by carbonizing the resin from the tree by a method similar to that used in the Mediterranean region. This method of producing carbon black was discovered in many areas. For example, during the Mayan civilization in Mexico, black pigment was made by carbonizing resin from the chacak tree. (In the Mediterranean region, carbon from cedar resin was added to cedar oil to make ink for writing on papyrus.)

These notable achievements of antiquity in developing processes for obtaining oleoresinous products from conifers and using them to produce varnishes, lacquers, paints, and inks remained the state of the arts for many centuries. Except for the eclectic writings of Theophrastus (372–287 B.C.), literature devoted to these processes suffered from historic amnesia well into the Renaissance.

2.1.10. First Varnish

Varnish based on solutions of amber were used as early as 250 B.C., but the formulations were not documented until a monk named Theophilus described the production of an oil varnish by dissolving resin in hot oil. Amber was one of the resins used at that time.* Subsequently, natural resins were obtained from trees in tropical regions. The resins were classified as ancient or fossil, semifossil and recent, depending on when they were separated from the trees. These resins were also named after their geographic source, such as Manila, Batu Dammer, Congo, and Kauri.

East Indians used shellac and sealing wax, incorporating the pigments in lac sticks and applying them to pieces of wood on a lathe. As the lathe turned, the friction created sufficient heat to melt the shellac and thereby apply a coating to the wood. It is interesting to note that the art of applying paint by brush was known at that early date, but the manufacturing methods of these brushes is unknown.

2.1.11. Mayan Civilization

The Mayan civilization of Mexico and Central America relied on natural organic pigments for its colors, which are very interesting, since they do not appear to be duplicated in other parts of the world. Whites were made by mixing lime with the juice of the chichebe plant; natural earths (iron oxides)

---

*The name varnish is derived from vernix, which is the Latin word for amber.

were used for browns and reds; the finest reds were made from splinters of heartwood; and blues were made from plants containing aniline compounds. They also had an inorganic blue called Mayan Blue. Yellows were made from the fruit of the achiote, or from chips of the fustics. Brushes were made from either bird plumes or hair. Many of the Mayan paints had good durability.

2.1.12. Medieval Europe

In Medieval Europe (Stillman, 1960), the art of paint, as well as the art of manufacturing paint was progressing. There is much literature on the subject of making various types of coatings and the ingredients used, but it is difficult for the uninitiated to determine just what is meant by certain raw materials whose names are to a great extent unknown today. However, it is known that the advent of oil as an ingredient in manufacturing varnish occurred sometime around the sixth century. Still the preferred medium for manufacturing paint was egg albumin.

2.1.13. Renaissance

Greater interest was shown in oils about the time of the Renaissance (Kranzberg et al., 1967). During this period, the artist was not only the painter but also was the paint manufacturer. Since the artist required only a small amount of paint, batches were only as large as needed for a particular project, and an ounce of paint might be a normal batch, while a really large mural could consume a 6-ounce batch. Manufacturing equipment, while crude, was still being used as lab equipment up to World War I, and apothecaries used the same mortar and pestle even in the early part of this century.

Treaties written during the fifteenth century (some are preserved in the Vatican) describe rosin and sandarac and preparing sandarac and mastic varnishes containing linseed oil. These varnishes were often used to coat armor, crossbows, and other weapons. An unidentified Flemish artist in the fifteenth century was able to produce paintings with unusual durability, but no one is certain just how he accomplished this feat. Speculation is that he painted with a tempera of an egg yolk emulsion and then coated the painting with a glaze made with an oil containing varnish. Many of the early artists were able to manufacture coatings of remarkable qualities; one of the most noted with Antonio Stradivari, who coated his magnificent violins with a finish that is unknown to this day.

As time passed, more varnishes were made from a variety of resins, at least one oil (linseed) and possibly others, and soon thinners were incorporated to allow the painter to use varnishes at ambient temperatures. Again, the largest batch was 6 ounces. Around the middle of the seventeenth century, it became apparent that driers had a place in paint. Such materials had been suggested as early as the second century, but they were not accepted until a millennium later. Bleaching oils was known, but little practiced, and oils were usually purified by boiling them in water.

Until the late 1600s, very few recipes for paint or varnish were ever recorded. Some were written for various grinding media, but by and large, everything was kept secret by the families that made and used the paints. Some of the old varnish formulae that have been published are intriguing, since they list materials seldom heard of today.

2.1.14. Early America

The first settlers lived in log cabins, which precluded the use of paint. As their building techniques improved and industrialization began, they learned to saw logs into boards and build primitive, unpainted conventional homes. As tastes changed, people began to decorate their homes, first on the inside and then on the outside. Prior to the seventeenth century, the use of paint on houses was mainly restricted to the wealthy. Each painter bought his/her own raw materials, since there were no stores that sold finished paint, and made his/her own paint according to a personal formula.

Watin, in 1773, was the first to detail the technical preparation of paints and varnishes. Copals and ambers were the resins of choice for varnishes at the time of the Revolution. Slowly but surely, varnish factories began to operate throughout Europe, and this may be termed the real beginning of the paint industry. Paint manufacturing itself progressed from a mortar and pestle type of operation to a rough stone trough with a ball muller enabling much larger batches of paint to be made with ease.

The Boston stone, a prime example of this device, was imported from England. In 1727, Thomas Child produced two barrels of oleoresinous paint daily in Boston using "ye old paint stone." The Federation of Societies for Coatings Technology has preserved the stone, and it is still visible as a building stone in downtown Boston.

2.1.15. Oleoresins

Varnish is defined as a composition of resinous matter in a volatile solvent, and conifers have been producing varnish compositions for the past 200 million years. Widely distributed over the earth, there are over 500 species of conifers that exude oleoresin (also called gum) consisting chiefly of a mixture of resin acids and turpentine.

2.1.15.1. Early Process for Obtaining Oleoresin

When burning conifer wood for cooking and making pottery, our ancestors observed oleoresin flowing out of the wood. Using a process similar to the one used in making charcoal, they built a stone furnace fitted with a conduit for the tar and pitch to flow out, taking care to avoid flames on the heated wood. The oil (turpentine) was recovered by placing flocks of loose wool or fleecy sheepskin above the heated wood to absorb the vapors. The oil (cedar oil, a turpentine), called cerium in the Near East, was obtained by wringing out the sheepskin.

Our ancestors learned to obtain oleoresin from conifers by making an incision with a stone axe in the living tree. By cutting deeply into the tree to form a box into which the oleoresin flowed, they were able to gather and transfer the exudate into clay pots. The exudate was cooked in the open clay pot until only a thick pitch remained. The oily vapors that rose from the boiling exudate were absorbed by a sheepskin placed over the pot and later wrung out to recover the oil. Thus, by using a crude distillation process, our early ancestors were able to separate oleoresin into three products: turpentine, tar, and pitch.

2.1.15.2. Using Oleoresin Products

In the eastern part of the Mediterranean where cedars abounded, Phoenicians were able to build seaworthy boats with cedar planks and use the tar and pitch from cedars to caulk their boats. This occurred long before Noah was instructed to caulk the ark "within and without with pitch." The Phoenician caulking art was diffused to Egypt where pitches and balsams were used as protective coatings for their boats. Egypt obtained cedars from Lebanon timbers, and tar, pitch, and cedar oil from Phoenicia for making and caulking their boats; cedar oil was also used to embalm the dead. By about 2500 B.C., Egyptians had developed varnishes and paints based on cedar oil and applied them to buildings, sculpture, and coffins for their mummies.

The Persian name for cedar oil was terpentin or termentin, which became terebinthine in Greece, terebentine in France, turpentyne in England, and turpentine in the United States. In ancient Egypt, amber dissolved in cedar oil was used as a varnish. This fossilized resin was called Bernice, which then became Verenice, Vernix, and in twelfth-century Medieval Latin to varnish. In Greece, Thales called it Electron (the property he observed was static electricity). Amber was first referred to as a fossilized resin from an extinct species of conifer in 1767 by Friedrick Samuel Bock in *Attempt at a Natural History of Prussian Amber*.

The production of tar, pitch, and oil in Greece from the Aleppo pine also predates biblical times. The Greeks named the crude exudate colophony for its geographical source, Colophon, in Asia Minor. In the literature on terpenoid colophony is a synonym for rosin.

2.1.16. Middle Ages

An oil varnish, prepared by dissolving molten resin in hot linseed oil was introduced in eleventh-century Europe. During the Renaissance, artists developed their own paints by using different pigments with rosin and sandarac in linseed oil. Leonardo da Vinci (1452–1519) used resin-coated pigments, and Rembrandt (1606–1669) used oleoresinous vehicles mixed with amber varnish. During the Renaissance, venic turpentine (chiefly pinene) was found in artists' wares. In the thirteenth century, turpentine produced in France from the oleoresin of conifers was a vehicle for paints and varnishes increasingly employed as a protective coating for buildings in Europe.

2.1.17. Naval Stores in the New World

From antiquity to well into the nineteenth century, the primary use of oleoresins was in tar and pitch for caulking wooden ships (Gerry, 1935). The centers of tar and pitch production were the Scandinavian and Baltic countries. England's need for these products for its navy and merchant ships prompted the 1607 establishment of a settlement at Jamestown, Virginia, where these materials were available. By 1608, the colony was producing tar and pitch for England. The term naval stores was introduced in England for these products for caulking sailing vessels, waterproofing tarpaulins, and treating rope. English sailors who worked with the tar and pitch became known as Jack Tars (Kirk-Othmer, 1970). North Carolina, which became the primary naval stores producer in 1665, was called the tarheel colony. Naval stores production of crude gum and the tar-burning process for producing tar and pitch was the first U.S. industry. Up to the twentieth century, the methods for obtaining oleoresin and producing pitch, tar, spirits of turpentine, and rosin changed very little from those used in antiquity.

By midnineteenth century, iron boats began to displace wooden boats. This prompted naval stores producers to separate turpentine from oleoresin to meet the growing demand for a paint and varnish solvent. In the 20th century, the demand for rosin in paint and varnish was so great that its production reached 8.7 million pounds in 1900 and over 10 million pounds in 1910, the highest in the history of gum naval stores. Then wood rosin dominated the supply and since the 1970's tall oil became the major source for turpentine and rosin (Drew, 1978).

2.2. PAINT IN THE NINETEENTH CENTURY

2.2.1. Coatings Industry

Although coatings technology was first described by Watin (Seymour, 1990) in his 1773 book, the first varnish factory was established in England in 1790. Shortly afterward, factories were established in several European nations and in the United States. Copal and amber continued to be the principal varnish resins and turpentine the thinner.

A major advance in coatings technology was made in 1880 by Henry Alden Sherwin, who with Edward Williams introduced the world's first ready-mixed paint. The Sherwin–Williams Company was formed in Cleveland, Ohio, to market the product. Professional and amateur painters thereafter could abandon the onerous task of combining white lead, linseed oil, turpentine, and colored pigments.

2.2.2. Early Chemical Knowledge of Paints and Coatings

Until chemistry became a science in the nineteenth century, the history of terpenoids in coatings was not one of steady transformations and improvements but rather a series of uneven lurches and trade-offs. The composition of oleoresin was a mystery except for the art of separating it into tar, pitch, and turpentine. Practically nothing was known about terpenes, resin acids, and their derivatives. Rosin and other natural resins, *e.g.*, amber, copal, dauri, and congo, were used to manufacture varnishes through the eighteenth and nineteenth centuries by cooking them plus a drier with linseed oil (Mattiello, 1941–1945).

Terpene chemistry could not blossom before organic chemistry became a discipline of chemistry. The earliest elementary analysis of turpentine was made in 1818, showing an empirical formula of C_5H_8, which became known as the isoprene unit. From 1852 to 1863, Marcellin Berthelot (1827–1907) characterized camphene and pinene as constituents of turpentine. The word terpene was introduced by Kekule in his 1866 textbook. Wallach began his extensive studies on the characterization of terpenes in 1884. Until high-plate distillation columns and sophisticated analytic spectroscopy and chromatography were available in the 1940s, terpene chemists confronted difficulties in separating the closely boiling terpenes in turpentine and other essential oils and resolving constitutional problems, since many terpenes undergo structural rearrangement under relatively mild conditions. Resin acid chemistry was even more difficult and complex than that of the terpenes in turpentine.

2.2.3. Rosin Esters in Coatings

Since Maly first esterified rosin in 1865, practically every conceivable ester of rosin has been prepared, including the glycerol ester, called ester gum, introduced in 1900 for lacquers and printing inks. The methyl ester of hydrogenated rosin is used as a plasticizer in lacquers. Pentaerythritol esters, unmodified and modified with maleic anhydride and phenol-formaldehyde, are used in varnishes. In 1937, the coatings industry used 20 million pounds of phenolic resins modified with ester gum; in 1987, 66 million pounds were used. Phenolic resins modified with up to 80% ester gum were called 4-hour enamels. Because these enamels possess excellent resistance to hot water and alkali, ester gum replaced damar, copal, and kauri gums in paints, lacquers, and enamels in combination with tung oil.

2.2.4. Phenolic Resins in Coatings

After ester gum, the first synthetic resin of importance to the coatings industry, phenolic resins were introduced in the early 1900s as a substitute for shellac. Phenol-formaldehyde, first prepared in the 1850s, was introduced in the 1920s for varnishes, paints, and lacquers by fusing it with rosin to make it oil soluble.

2.2.5. Terenpid Alkyd Resins in Coatings

A major contribution of naval stores products to the coatings industry occurred in the 1920s and 1930s with the production of alkyd resins from the reaction of polyhydric alcohols, *viz.*, glycerol and pentaerythritol with rosin-maleic and terpene-maleic Diels-Alder condensation products. The name alkyd was coined by Kienle (Simonsen *et al.*, 1947–1952) in 1927.

Rosin-maleic glyceride yields a lighter colored and nonyellowing varnish compared to phenolic resin varnishes, and it is more compatible with nitrocellulose in lacquers. Reaction of the rosin-maleic adduct with a dihydric alcohol followed by heating with a monohydric alcohol ester of an unsaturated resin yields an alkyd with a very low acid number for use with vinyl polymers in coating applications. Although terpene-maleic adducts are more expensive than phthalic anhydride, its glyceride alkyd has the advantage of greater solubility in cheaper solvents and better compatibility with superior adhesive and flowing properties. With the advent of nitrocellulose lacquers, *e.g.*, Duco®, in the early 1920s for finishing automobiles and later used for furniture, camphor and ester gum were used as plasticizers allowing the lacquer to be applied by dipping or spraying. Rosin-maleic adduct (Skolnnik, 1959) esterified with a polyol yields a hard resin, which has been used with nitrocellulose lacquer for finishing furniture.

2.2.6. Rosin Salts in Coatings

Rosin salts of polyvalent metals, *e.g.*, Ca, Zn, Pb, and Mn, have been used as driers for paint and varnish and in printing inks. Six million pounds of rosin salts were produced in 1980, mostly from tall oil rosin; the NH_4^+ and Na^+ resinates have been used in emulsion paints.

Turpentine (Skolnik, 1968, 1983) has long been used as a thinner or solvent for paints and varnishes. Its solvent and wetting properties are superior to those of straight petroleum solvents. Nevertheless, since World War II, turpentine has been steadily replaced by petroleum solvents and eliminated by the growing popularity of water-based paints. Gum naval stores production has decreased markedly since 1930 when wood naval stores took over. Then in the 1950s, tall oil, a by-product of Kraft paper manufacturing became the dominant source for turpentine and rosin (Skolnik, 1983). Since the mid-1970s, the total naval stores market in coatings has been declining. Only ester gum, other synthetic resins of rosin and terpenes (Zinkel, 1975), and a number of rosin and terpene derivatives have experienced a relatively steady market in coatings.

2.2.7. Waterborne Coatings

As a result of regulations limiting the amount of volatile organic solvents in coatings, there has been a renewed interest in water-borne coatings, which were the earliest types of paints used. Early man used earth colorants as pigments and crushed berries, animal blood, egg whites, and sap from dandelions, milkweed, and trees as adhesives in these crude paints.

Paintings of the grand bison in Altamira, Spain, and the Chinese horse in Lascaux, France, are believed to be 15,000 years old. The Obiri rock paintings in Arnhem land in Northern Australia also date back to prehistoric times. Some 5000 years ago, Egyptians improved their coatings by adding other adhesives, such as casein. The term distemper from the Latin temperare, meaning to mix, was used to describe these waterborne paints.

2.2.8. Pigments

The principal pigments used by the Palaeolithic artisians were charred wood (black), chalk (white), and iron and manganese oxides for red and yellow colors. About three or four thousand years ago, Egyptians supplemented these basic pigments with lapis lazuli (blue), azurite (green), ochres (red and yellow), orpiment (yellow), malachite (green), gypsum (white), and lamp black.

Egyptians also developed such vegetable colorants as those from the madder root. The ancient Romans used red lead as a pigment in coatings, but the first synthetic pigments were Egyptian blue and white lead (cerussa). Egyptian blue was produced by the calcination of a mixture of lime, alumina, silica, soda ash, and copper oxide at least 10 thousand years ago. White lead was made by Pliny by the reaction from vinegar on lead sheets in the atmosphere over 2500 years ago.

Other classic white pigments were zinc oxide, zinc sulfide, lithophone, and basic white lead sulfate, which were introduced in 1770, 1783, 1847, and 1855, respectively. Titanium dioxide, produced from black ilmenite in 1924, became the major white pigment, and it is the world's most widely used pigment.

2.2.9. Oleoresinous Paints

The term paint is usually applied to a coating consisting of pigments dispersed in a drying oil, such as linseed oil, but through misuse, this term has been expanded to include many other coatings, including water-based coatings.

There is some evidence that drying oils were used in ancient Egypt, and more definite reports indicate that linseed oil was used for coatings in the fourth century A.D. Linseed oil, obtained from the seeds of flax (*Lininum usitatissimum*), was the first vegetable oil binder used in making paints. This oil is a glycerol ester of unsaturated acids, such as linoleic and oleic acids. Linseed oil forms a film when exposed to the atmosphere, and its hardening rate is accelerated when white lead is mixed with the oil. It has been suggested that a trace of free acid in the oil reacts with the lead salts to produce an oil soluble salt, which is called a drier or siccative.

Presbyter Theophilus (twelfth century), also called Rugerus, in cooperation with another monk, called Roger, wrote "De Diversis Artibus," which described painting and other practical arts used in church decoration. Neither of the coauthors showed any scientific talent, but they did record unpublished recipes, which incorporated drying oils as linseed, that had been used for centuries. Detailed recipes for producing oleoresinous paints were recorded by Theophilus.

Linseed oil continues to account for more than 50% of the drying oils used in paints, but its share of the market, which was over 90% in 1900, has decreased due to the use of soybean oil, tung oil, dehydrated castor oil, oiticica oil, and fish oil. The hardening or drying of these glycerol esters in air depends on the extent of unsaturation, and it is accelerated by the addition of driers.

In spite of their empirical development and lack of understanding of the drying process, oleoresinous formulations became the standard in producing paints, putty, oil cloth, linoleum, artist's color, and printing ink until the early 1920s when these were replaced, to some extent, by synthetic polymers.

2.2.10. Driers

Lead salts of organic acids, the original driers, function as polymerization catalysts throughout the entire paint film depth; in contrast, cobalt salts function as surface driers. Hence, until recently, combinations of lead and cobalt driers have been used to achieve a uniform drying rate. Manganese and zirconium salts may be used in place of lead. Many organic acids have been used to form these heavy metal salts, but the most widely used salts are naphthenates, which are based on naphthionic acid, a residue of petroleum resinous. Other acids used to produce driers are octoic, tallic, rosin, and linolenic acids.

2.2.11. Drying Paints

It has been recognized for many centuries that film formation from oleoresinous paints depends on atmospheric exposure of the unsaturated oils. It is now known that a crosslinking between the polymer chains occurs and an insoluble polymer network is produced after oxygen is absorbed in this drying reaction.

Drying nonconjugated acids involves the formation of hydroperoxide groups on the allylic carbons, i.e., carbon atoms of the methylene groups adjacent to the double bonds. Polymerization occurs via a radical chain mechanism, called autoxidation. This theory also applies to conjugated acids, such as linolenates, but the drying reaction is much faster with these polyunsaturated oils.

2.2.12. Cellulose Nitrate (Pyroxylene)

Solutions of shellac in ethanol and sumac sap were used as coatings centuries ago, but the first lacquer from a man-made polymer was not available until the middle of the nineteenth century. Broconnet nitrated starch in 1833, and Pelouze nitrated cellulose 5 years later. The names of xyloidine and pyroxylene, respectively, were used by the inventors to describe their products.

In 1846, Schonbein improved the nitration process by using a mixture of nitric and sulfuric acids; he patented this cellulose trinitrate, which he called guncotton. Schonbein observed that guncotton was soluble in a 50:50 mixture of ethanol and ethyl ether, but credit is usually given to Maynard, who announced the availability of collodion as a waterproof coating for wounds in 1847. In

1882, Stephens used amyl acetate as a solvent for cellulose nitrate, and Wilson and Green patented pyroxylene coatings for carriages and automobiles in 1884.

The availability of huge stockpiles of surplus guncotton after World War I catalyzed peacetime uses for this polymer. In 1923, cellulose nitrate (CN) lacquers, erroneously called nitrocellulose lacquers, were used as automobile finishes under the trade name of Duco. After the introduction of competitive resins for automotive coatings, catalyzed lacquer and a multicolored CN were developed in the 1950s. Super lacquers based on cellulose nitrate-ioscyanate resins were introduced in the 1960s when CN continued to hold second place, next to alkyds, among industrial finishes. Alkyd–CN blends and CN blends with copolymers of vinyl chloride and vinyl acetate are also used as commercial coatings.

2.3. PAINT IN THE TWENTIETH CENTURY

2.3.1. Phenolic Resins

Phenolic resins, which are noncrystalline and lack a precise melting point, were produced and discarded as undesirable "goos and gunks" by leading organic chemists in the late nineteenth century. As a result, many of the first- and second-generation students of these professors avoided the resins and other polymers, concentrating their efforts instead on crystallizable and distillable compounds.

In 1872, Bayer condensed phenol with methylol, and at the suggestion of Fischer, Kleeberg repeated this experiment, using formaldehyde instead of methylol in 1891. Smith patented phenolic resins for use as electrical insulators in 1899, and Swinburne produced these products commercially in 1904.

Baekeland, a visitor from Belgium, who decided to remain in the United States, was aware of the importance of functionality and the mistakes made by his predecessors in investigating phenolic resins. Fortunately, he had been awarded enough money by Eastman for his Velox photographic paper patent to enable him to establish his own laboratory and choose his research projects.

In 1907, Baekeland produced an ethanol-soluble novolac, which was used as a substitute for shellac, from the condensation of an excess of phenol with formaldehyde under acid conditions. In 1908, he produced resole prepolymers from the condensation of phenol and formaldehyde under mild alkaline conditions. These prepolymers, called A-stage, could be converted into infusible C-stage insoluble castings or coatings in the presence of strong acids. Nevertheless, neither the novolac or resole resins could be substituted for natural resins in oleoresinous varnishes.

Albert however heated the phenolic resin (PF) with a large excess of rosin and produced a product that could be used with tung oil to yield 4-hour enamels. In 1928, Turkington patented oil-soluble varnishes produced from the alkaline condensation of formaldehyde with para-substituted phenols, such a p-phenylphenol.

2.3.2. Amino Coatings

Tollens described resins produced from the condensation of urea and formaldehyde in 1884, and John patented this polymer (UF) in 1918. In 1936, Henkel patented comparable resins (MF) based on melamine instead of urea. Both resins, which were referred to as amino resins, are insoluble in common solvents and used as additives for curing other coatings. For example, the addition of MF reduces the curing time of alkyd resins by 50%. Soluble amino resins, which can be used as coatings, are produced when modified by etherification with butanol.

2.3.3. Vinyl Chloride Polymers and Related Resins

Regnault described polyvinyl chloride (PVC) in 1835, but since no solvents were available for this polymer, no PVC coatings were produced. In 1920, Reid of Union Carbide and Voss and Dickhauser of I. G. Farbenindustrie filed for patents on a more soluble copolymer of vinyl chloride and vinyl acetate. This copolymer, produced under the trade name of Vinylite, was used to coat beer cans in 1936.

The versatility of this copolymer, copolymers of vinyl chloride and vinylidene chloride, and PVC increased in 1944 with the development of plastisols by suspending a resin produced by emulsion polymerization in a liquid plasticizer and then fusing the mixture at elevated temperatures.

While polyvinyl alcohol (PVAL), obtained from the hydrolysis of polyvinyl acetate (PVAC), is insoluble in organic solvents, the acetal produced from the condensation of butyraldehyde and PVAL is soluble and used as a base for "wash primers" or metal conditioners.

Polyvinyl acetate (PVAC), patented by Klatte and Rollet in 1914, is used as an adhesive and the major resinous component of a widely used water-borne coating.

2.3.4. Acrylic Esters

While acrylic acid was polymerized in 1847, its esters remained laboratory curiosities until the 1900s when Rohm wrote his Ph.D. dissertation on acrylic esters. He continued investigating these products for several decades. In 1956, General Motors Company replaced some of its cellulosic automotive finishes with acrylic coatings. Some of these were thermoset by incorporating melamine resins or introducing additional functionality in the acrylic monomers.

2.3.5. Silicones

Polysiloxane was synthesized by Kipping in the early 1900s. Since he believed that these polymers were ketones, he called them silicones. While he was pessimistic about their commercial use, these water- and temperature-resistant coatings were commercialized in the 1930s by Rochow of General Electric Company, McGregor of Union Carbide, and Bass of Dow Corning.

2.3.6. Epoxy Resins

Ethyoxyline resins, now called epoxy resins, were patented by Schlack in 1939. Many of the prepolymers were versatile resins produced from the condensation of a diphenol (bisphenol A) and epichlorohydrin. Since these resins possess terminal oxirane (epoxy) groups, they can be crosslinked in reactions with polyamines at ordinary temperatures. These resins may also be crosslinked by esterifying the pendant hydroxyl groups by cyclic anydrides at elevated temperatures.

2.3.7. Polyurethanes

The original resins that Bayer produced in 1937 by condensing aliphatic diisocyanates and diols were elastomers and foams. However, both one- and two-component polyurethane coatings are produced by heating phenol-capped isocyanates to remove the phenol and permit the isocyanate group to react with diols present in the coating system.

2.3.8. Polyesters

Since some of the first synthetic organic compounds were esters produced by condensing monofunctional alcohols, such as ethanol, and monofunctional organic acids, such as acetic acid, it is not surprising that one of the first synthetic polymers was polyglyceryl tartrate. This compound was synthesized from disfunction reactants by Berzelius in 1847. Over a half-century later, Watson Smith produced thermosetting polyesters (Glyptals) by condensing glycerol and phthalic anhydride in 1901.

During the following half-century, polymer chemists learned that curing or drying oleoresinous paints in the presence of heavy metal salts ("driers") and air involved forming hydroperoxides on the carbon atoms adjacent to the ethylenic bonds in the unsaturated oils. In the accepted mechanism, the carbon-carbon double bond shifts to a conjugated configuration, and the peroxide is transferred to another monoallylic carbon atom. Polymerization proceeds via a radical chain mechanism to produce a crosslinked insoluble film.

2.3.9. Alkyds

The first polyester resin was produced by Berzelius from the condensation of glycerol tartrate in 1847. W. Smith made Glyptal coatings in 1901 from the controlled esterification of glycerol by phthalic anhydride. This technology was improved by Friedburg a few years later and by Kienle in 1921.

Kienle recognized that the condensation of difunctional reactants produced linear polymers and trifunctional reactants produced crosslinked, infusible network polymers. Accordingly, he used a reaction of ethylene glycol and phthalic anhydride in the presence of drying oils to obtain linear polymers that could undergo autoxidation polymerization like that described for oleoresinous paints. Kienle used syllables from the reactants *al*cohol and *a*cid to coin the word alkyd.

In 1931, Kienle used Glyptal and oleoresinous paint technology to produce superior polymers that now account for about one-half of all coating resins. He obtained air-curable resins that were more flexible than Glyptals by producing oil-modified polyester resins for which he coined the acronym, alkyd, from the *al*cohol and *a*cid reactants used.

Kienle also used Baekeland's concept of functionality, which was quantified by Carothers. When the functionality of the reactants is greater than 2, crosslinking or "bodying" occurs as the glycerol, phthalic anhydride, and unsaturated oil are heated. Kienle used a reaction of unsaturated oil with glycerol at 446–554°F (230–290°C) in the presence of litharge to produce a monoglyceride, which was then esterified in a reaction with phthalic anhydride.

Kienle classified his alkyd resins as short, medium, and long oils on the basis of possessing an oil content of less than 40%, less than 50%, and over 50% oil, respectively. The Kienle patent was declared invalid by the U.S. Patent Office in 1935, but the name alkyd is still used to describe the major coating resin used worldwide.

Medium oil alkyds are more versatile than short or long oil alkyds. Most alkyd resins are produced by the fatty acid process in which the glycerol, dicarboxylic acid and unsaturated acid are condensed at 392–446°F (200–

230°C). Alkyd coatings have superior gloss, adhesion, hardness, and chemical resistance, and they dry faster than oleoresinous coatings.

2.3.9.1. Alkyd Blends

Selected alkyd resins are compatible with cellulose nitrate, amino resins, phenolic resins, epoxy resins, silicones, acrylics, chlorinated rubber, and styrene. When added to cellulose nitrate, short-oil alkyds improve gloss, adhesion, and flexibility of these commercial coatings. The chemical resistance of short-oil alkyds is improved when they are reacted with amino resins. Alkyds also react with phenolics to produce chroman-type alkali-resistant coatings.

The properties of alkyds and epoxy resins are upgraded when both resins are present in a coatings mixture. Heat resistance, durability, and gloss of alkyds are also upgraded when they are reacted with silicones. It is customary to add alkyd resins to improve the durability of highway-marking paints based on chlorinated rubber. Long-oil alkyd resins are also added to latex house paints to improve their resistance to chalking.

2.3.9.2. Acrylated Alkyds

The degree of hardness of alkyd resins has been increased by blending them with low molecular weight acrylic resins. More homogenous resinous coatings have been obtained by blending acrylic acid copolymers with alkyds. These acrylated alkyds dry more rapidly than alkyd coatings, but they are not available commercially.

2.3.9.3. Styrenated Alkyds

The drying rate and hardness of alkyd coatings were improved in 1948 by Bhow and Payne, who added polystyrene to the alkyd resins. Xylene solutions of these styrenated alkyds were produced by several different techniques, but the polymerization of styrene in a solution of the alkyd was preferred.

In a recipe supplied by Payne, a solution of 4 moles of styrene and 3% benzoyl peroxide was slowly added over a 30-minute period to a xylene solution of dehydrated castor oil at 239°F (115°C). These solutions were then heated at 284°F (140°C) for 1 hour. The 6.5% unreacted styrene monomer was removed by vacuum distillation prior to traction with phthalic anhydride 449.6°F (232°C) for 80 minutes. However, in the most widely used method, styrene and the benzoyl peroxide are added to a previously prepared resin. Harrison, Tolby, and Redkamp maintained that unless long-oil alkyds were used, the styrenated alkyds were primarily blends of polystyrene and the alkyd resins.

Precipitation in methanol has shown that many commercially available styrenated alkyds are blends of the alkyd and polystyrene. An attempt to produce copolymers by adding styrene and benzoyl peroxide or di-tert-butyl peroxide to a commercial soy-linseed medium alkyd (Cargill Company Alkyd Resin Product No. 5150) in a nitrogen atmosphere, were also unsuccessful. However, blends containing at least 40% styrene were compatible with the alkyd resin, and the coatings were slightly hazy, fast drying, harder, and more chemical resistant than the alkyd resins component.

Graft styrene-alkyd copolymers were obtained when a solution of styrene and tert-butyl perbenzoate were added simultaneously to the medium oil alkyd at 284°F (120°C). The reaction was continued for 2.5 hours after adding styrene and initiator, but a small amount of unreacted styrene remained in the solution. Precipitation in methanol and characterization by FTIR spectroscopy and gel permeation chromatography (GPC) showed the presence of the graft copolymer rather than a mixture.

2.3.9.4. Alpha-Methyl Styrenated Alkyds

Attempts to produce a copolymer by heating alpha-methylstyrene and tert-butyl perbenzoate with an alkyd were unsuccessful. However, a graft copolymer was obtained with a 50:50 mixture of alpha-methylstyrene, styrene, and tert-butyl perbenzoate heated with the medium-oil alkyd resin.

2.3.9.5. Vinyl-Toluenenated Alkyds

Dow produces a mixture of meta (60) and para (40) isomers of methyl-styrene (VT), which it calls vinyltoluene (VT). Blends of VT and medium-oil alkyds, which are produced when VT is heated at 266°F (130°C) in the presence of di-tert-butyl peroxide, are compatible, but the films are hazy. However, graft copolymers are produced when tert-butyl perbenzonate is used as the catalyst at 248°F (120°C) instead of di-tert-butyl peroxide at 266°F (130°C). These graft copolymers are soluble in aliphatic hydrocarbon solvents, such as naphtha.

2.3.9.6. Para-Methylstyrenated Alkyds

A methylstyrene monomer, which is a 97% para and 3% meta isomer, was produced commercially by Mobil. This production facility was acquired by American Hoechst and sold to Deltech in 1988. The Para-Methylstyrenated (PMS) copolymerizes readily in the presence of di-tert-butyl peroxide and a medium-oil alkyd at 266°F (130°C), but the product is a compatible mixture of the two polymers that can be separated in methanol.

Coatings from these mixtures are unaffected by water, but they pit and discolor in ethyl acetate and crack in acetone. Their hardness, as measured by a Sward rocker, is much higher than the alkyd of a 40–60% blend of the alkyd and PMS.

The graft copolymer of PMS (40)-alkyd (60) has a Sward rocker harness of 46 versus 30 for the corresponding blend. Neither the copolymer nor the blend is adversely affected by water or methanol, but both bubble when exposed for 24 hours to ethyl acetate or xylene. These solutions contain about 1% residual monomer, and this monomer content can be reduced by vacuum distillation.

Styrenated alkyds produced by the copolymerization of the medium-oil alkyd (60) and styrene, vinyltoluene, or para-methylstyrene (40) in the presence of p-tert-butyl perbenzoate are harder than the unreacted alkyd. The coatings obtained from the graft copolymer are harder, more solvent resistant, and more ductile than coatings obtained from blends with similar ratios of polymers.

Coatings continue to be used at an unprecedented rate for both decoration and protection. Over 1 billion gallons of coatings are produced annually in the United States at a selling price of $10 billion.

2.4. MAJOR EVENTS IN PAINT TECHNOLOGY

Table 2.1 summarizes the major events in paint technology.

Table 2.1.   Chronology of Major Events in Paint Technology

Years	Event
25,000 B.C.	Painted drawings in caves by Neanderthal and Cro-Magnon man in Europe and Australia; natural pigments, such as minerals; and animal fat and egg yolk used as binders.
19,000 B.C.	Paint made from natural sources used by Egyptians.
15,000 B.C.	Water-based coatings used in Altamira, Spain, and Lascaux, France.
9000 B.C.	Painted drawings in North America prepared from natural pigments and binders, possibly painted by Caucasians.
2500 B.C.	Cedar oil was used by Egyptians for coffins, buildings, and sculptures.
	Terpentin or turmentin, the Persian name for cedar oil; became terebinthine in Greece, terebentine in France, turpentyne in England, and turpentine in the United States.

(*Continued*)

Table 2.1.   (*Continued*)

Years	Event
2500 B.C.	Amber (called Bernice) was dissolved in cedar oil and used as a varnish in Egypt; Bernice gradually became Verenice, Vernix, and in the twelfth-century A.D. Medieval Latin, to varnish.
1000 B.C.	Casein from milk used by Egyptians in water-based coatings; the term distemper from the Latin temperare, meaning to mix, was used to describe these water-based paints.
200 B.C.	Lacquers prepared from the exudation of the conifer *Rhus vernicera*, called the Urushi tree (varnish tree).
	Paint prepared from natural pigments and binders used by the American Indians for tepees and war paint.
100 B.C.	Greeks and Romans made paints; Romans prepared artificial pigments, including white lead, litharge, red lead, and yellow lead pigments.
200	Lake pigments made using woods, kermes, and madder; pitch used for sealing ship bottoms.
400	Linseed oil—the first vegetable oil binder used for making paints in Egypt.
	Laquers called Japanese Laquer prepared in Japan from the conifer *Rhus vernicera* found earlier in China.
1000	Shellac (called Indian Amber by Pliny) was one of the few resins obtained from an insect (*Laccifer lacca*); lac sticks made for coating in India and Thailand.
1050s	Oleoresinous paint recipes recorded by Theophilus.
1600	Varnishes, linseed oil, and thinners used in Europe.
1608	Tar and pitch produced at Virginia for export to English naval stores; used for waterproofing tarpaulins, caulking sailing vessels, and treating rope.
1650	Driers used to accelerate curing oils used in paints.
1727	Two barrels of oleoresinous paint produced daily in Boston, Massachusetts, by Thomas Child using "ye old paint stone" imported from England.
1773	Copals and ambers used to make varnishes in North American colonies; Watin first described technical preparation of paints and varnishes.
1832	Cellulose nitrated by Braconnot.
1835	Polyvinylchloride described by Regnault.
1838	Cellulose nitrate prepared from paper by Pelouze.
1840	Cellulose nitrate (nitrocellulose) now officially accredited to Braconnot and Pelouze.
1845	Schonbein invented cellulose trinitrate; called it guncotton (patented in 1846).
1847	Maynard used collodion (guncotton) as a waterproof coating for wounds.
	Borzelius synthesized first polyester resin.
	Polymerization of acrylic acid.

(*Continued*)

Table 2.1. (*Continued*)

Years	Event
1865	Rosins first esterified by Maly.
1872	Bayer synthesized first phenolic resin.
1880	First ready-mixed paint produced by Henry Alden Sherwin and Edward Williams; Sherwin-Williams Company formed in Cleveland, Ohio, to market the product.
1882	Stephens used amyl acetate as a solvent for cellulose nitrate.
1884	Wilson and Green patented pyroxylene coatings for carriages and automobiles.
	Amino-resins from condensation of urea and formaldehyde first produced by Tollens
1900s	Kippling synthesized first polysiloxane.
	Rohm wrote Ph.D. dissertation on polymerization of acrylic esters.
1901	Glyptal (thermosetting polyester) resins prepared by Smith from the condensation of glycerol and phthalic anhydride.
1907	Baekeland produced first ethanol-soluble novalac; used as a substitute for shellac.
1918	Amino-resins from urea and formaldehyde; patent application filed by John.
1920	Soluble copolymers of vinyl chloride and vinyl acetate; patent applications filed by Reid of Union Carbide, and Voss and Dickahauer of I. G. Farbenindustrie.
1921	Polyester resins improved by Kienle; applied unsuccessfully for patent for alkyd resins, acronym from alcohol and acid reactants used.
1923	Cellulose nitrate (trade name Duco) lacquers used in automobile finishes.
1927	Alkyd name officially coined by Kienle for ester oils.
1928	Oil-soluble varnishes patented by Turkington; produced by alkaline condensation with para-substituted phenols.
1931	Glyptal and oleoresinous paint technology (oil-modified polyester resins) used by Kienle to produce superior polymers that were air-curable and more flexible than Glyptals; became known as alkyd resins.
1935	Ethyl cellulose first went into commercial production.
1937	Polyurethane resins originally synthesized by Bayer.
1939	Ethyoxyline resins patented by Schlack; became known as epoxy resins.
1944	Polyvinylchloride plastisols were developed.
1946	First styrene-butadiene latex commercialized for paper-coating application; first patents on pigmented paper coating issued in 1954 to S. D. Warren.
1948	First styrene-butadiene latex sold for architectural-coating applications; prototype styrene-butadiene latex copolymer known as Dow latex 546.
1950	Multicolored cellulose nitrates developed.

(*Continued*)

Table 2.1. (*Continued*)

Years	Event
1953	Powder coatings developed by Gemmer of Farbwereke Hoechst; granted a patent on the fluidized bed method of applying coatings using a powdered polymer. Ethyl acrylate and methyl methacrylate copolymer introduced as binders in house paints; the product was called Rhoplex® AC-33.
1954	Cellulose acetate butyrate resin introduced for general purpose uses.
1956	Acrylic esters replaced some cellulosic automobile coatings by General Motors.
1960	Cellulose nitrate-isocyanate resins introduced. Alkyd-cellulose nitrate blends with copolymers of vinyl chloride and vinyl acetate used.
	Electrospray gun for applying powder coatings developed by Fraser and Point.
1963	Cellulose nitrate lacquers ranked second among all industrial finishes.
1970	Powder coatings used in Japan on automobiles and to a lesser degree in the United States; powder coatings eliminated volatile organic compounds.
1980s	Powder coatings used extensively on appliances and automobiles as a substitute for enamels, thereby eliminating volatile organic compounds.
1990s	Powder coatings expected to grow to 286 million pounds in North America and 507 pounds annually in Europe.

3

Fundamentals of Paint

3.1. DESCRIPTION, DEFINITION, AND CLASSIFICATION OF PAINT

3.1.1. Description

This section surveys paint technology from the user's point of view. Emphasis is on the fundamentals of paint technology and the practical use of paint. Paint ingredients are discussed so that the reader understands why paint behaves as it does and why it must be used in certain ways. The failure modes of paint are discussed, which explains how an area in or outside of a building can become contaminated with lead from paint.

Like all technologies, paint technology has its own jargon. The terms paint and coatings are sometimes used interchangeably although there are differences. Paint is an older term, used before the 1940s. After that, new sophisticated and synthesized materials developed for automobiles and aircraft were called coatings to distinguish them from vegetable-oil-based materials. The majority of lead-based paint discussed in this book consists of oil and alkyd resins.

Paint differs from other engineering materials because its successful use requires a proper mixture of science and art. Paint also has inseparable functions: decorative, protective, and specialized functions are all present simultaneously. Once the coating system has dried or cured, the separate layers cannot be cleanly separated to examine them. As a result, effects on various layers of different factors that have influenced their formation and performance cannot be studied individually.

Lead-based paint is composed of several different materials, and not all of the materials contain lead. Paint has been developed for many different purposes using various formulations. Below a certain level of lead content, dried paint is not considered a hazardous material (see Chapter 6). Lead sulfate pigment is considered less toxic than lead carbonate and other pigments. Not all lead in paint is used for pigmentation.

3.1.2. Definition

We define paint as a decorative, protective or otherwise functional coating applied to a substrate, which may be another coat of paint. The following are some common terms and descriptions of paint.

- Dopant (D. *doop*, adj.). Any thick liquid or pasty preparation used in preparing a surface; any varnishlike material for water-proofing surfaces.
- Paint (M.E., *peint*, n.). A substance composed of a solid coloring matter suspended in a liquid medium.
- Coating (M.E., *cote*, n.). A layer of any substance spread over a surface; modern synthesized materials, such as polyurethane resins, that replace older paint materials.

3.1.3. Classification

Paints can be classified by many methods. One way of classifying paint is to group those with similar properties. For example, gloss paints have a shine like a mirror, which flat paints lack. Another way of classifying paints is to group those used in the same way or for the same purpose or type of application. For example, industrial finishes are applied to manufactured objects (automobiles, appliances, and furniture) before they are sold; trade sales paints (house paints, wall paints, and kitchen enamels) are applied to completed articles after they are sold.

3.2. IMPORTANT PROPERTIES

Certain general properties (*e.g.*, adhesion, ease of application, film integrity, and consistent quality) exist in all paints to assure their decorative, protective, and useful functions as an engineering material. Other more specific properties, such as gloss and color, are related to end use.

3.2.1. Adhesion

Coatings must adhere to the substrate to develop other properties. Even a coating designed to be removed later must adhere to the substrate until that time.

3.2.2. Ease of Application

Paints must be easy to apply, and to be practical engineering materials, there must be a minimum of lost time or additional expense when painting. Paints must also be applied to the substrate in the specific film thickness, dry in the specified time to the desired appearance, and possess the necessary specific properties.

3.2.3. Film Integrity

When properly applied, the cured or dried film of paint should have all the film properties advertised by the paint manufacturer. This is film integrity, or uniformity. There must be no weak spots (holidays) in the film caused by imperfect drying or curing.

3.2.4. Consistent Quality

To function as engineering materials, paints must be consistent in quality from can to can, batch to batch, and shipment to shipment. Color, viscosity, application properties, durability, *etc.*, must all comply with specifications.

3.2.5. Specific Properties

The following are specific properties that should be taken into account when specifying acceptable paint standards for a particular use.

- Kitchen enamels must resist grease, heat, and repeated cleaning.
- Stucco paints must resist water, alkali, and sunlight and permit the passage of water vapor.
- Swimming pool paints must resist pool chemicals, especially acids and chlorine; water; and sunlight.
- Exterior commercial aircraft finishes must resist ultraviolet degradation, erosion by air, loss of adhesion at high speeds, rapid temperature changes, chemical attack by hydraulic fluids, and film rupture.
- All these paints must be made in the specified color, gloss, etc. They must also have the necessary ease of application and other essential properties.

3.3. REQUIREMENTS FOR AN ACCEPTABLE PAINT JOB

Performance of the following five tasks must be done to obtain an acceptable paint job:

- preparing the surface correctly
- protecting the surface from water
- choosing the proper paint system
- using the correct techniques and tools
- following the correct drying or curing procedure.

3.3.1. Correct Surface Preparation

The essential property of paint is adhesion, and good adhesion requires proper surface preparation. Although correct surface preparation varies somewhat with the type of paint being applied, the following five factors affect all paint classes.

3.3.1.1. Surface Integrity

The surface must be knit together tightly enough to resist shrinkage of the curing paint film. If the surface is loose or crumbly, the paint system will lose adhesion and peel off.

3.3.1.2. Surface Cleanness

The surface must be free of any foreign material that will prevent the paint from flowing onto the surface and covering it. If the paint cannot flow, it will lack film continuity, and if it does not cover the surface, it will lack adhesion.

3.3.1.3. Surface Smoothness

The surface must be filled, sanded, *etc.*, to produce the required smoothness of the finished paint system (combined coats of paint applied). The paint system will faithfully reproduce the contours of the substrate unless a rough-textured paint has deliberately been chosen to hide surface imperfections. This reproduction of substrate contours by the paint system is termed photographing.

Sometimes good adhesion is difficult or impossible to obtain without minute surface roughness, which need not be pronounced enough to affect the surface smoothness of the completed paint system. Mechanical anchorage due to the "tooth," or anchor pattern of the substrate is required for paint systems subject to vibration, impact, and bending, particularly when the paint is subject to weathering; the automobile's paint system, for example, is subject to weathering.

3.3.1.4. Surface Porosity

The suction, or porosity, of the substrate must be reduced so that the coating is not absorbed by the substrate. If it is, film continuity is destroyed, and ease of application is impaired.

3.3.1.5. Substrate Protection

Some substrates, especially plastics, are attacked and irreparably damaged by solvents in otherwise desirable coatings. In such cases, the composition of first coat, whatever it is called, must be such that its solvents do not attack the substrate. The first coat, in turn, must not be attacked by solvents in subsequent coats.

3.3.2. Choosing the Proper Paint System

A paint system is a combination of coats of paint that

- perform the desired function or functions of paint: decoration, protection, specialized functions
- have all the essential properties of paint and the necessary specific properties for the specified use.

Paint systems are emphasized because a single coat of paint will not normally do everything that is required. Even when a single coat of paint is sufficient for repainting, we are still dealing with a paint system, because there are underlying coats of old paint. Failing to consider the old paint and the new paint as a combined total paint system could result in the following situations.

- The new coating may attack the old coating, as when lacquer is applied over oil paint. In this case, a protective sealer should be used first.
- The old paint may not resist the shrinkage of the new paint, causing "alligatoring." In this case, an undercoat should be used first.
- The old paint may be too porous to seal the surface for the new paint, thereby causing blotching, poor film continuity, and other film weaknesses. In this case, a sealer should be used first to seal the surface.
- The total paint system may be too thick, causing cracking from thermal expansion and contraction. In this case, the old paint must be removed.

The following operations performed by paint systems could never be performed by a single coat of paint.

3.3.2.1. Binding the Substrate

This operation provides the needed surface integrity. Applying a surface conditioner to a chalky stucco surface before applying an emulsion paint binds the substrate together.

3.3.2.2. Staining the Substrate

This operation provides color without film buildup.

3.3.2.3. Filling the Surface

This operation involves building up low spots in the surface by filling and sanding. Filling the surface prevents surface irregularities from photographing through the paint system.

3.3.2.4. Sealing the Substrate

This operation involves using a sealer or primer to reduce and even out substrate porosity.

3.3.2.5. Protecting and Decorating without Opacity or Color

This operation involves applying a clear coating, such as a clear varnish or an oil finish, to preserve the existing surface color or wood grain.

3.3.2.6. Changing the Surface to Improve Adhesion

This operation normally involves applying a primer. In extreme cases, as with some urethane coatings, two primers are needed, so that each coat will adhere to the preceding one and also provide good adhesion for the following coat.

3.3.2.7. Obtaining the Desired Color

This operation usually depends on the topcoat alone; however, some automobile finishes, antiquing finishes, and a few other finishes require two or more coats.

3.3.2.8. Obtaining the Desired Level of Gloss

This operation depends on the topcoat. Old paint or an improperly cured or formulated undercoat can affect the topcoat gloss.

3.3.2.9. Providing the Necessary Specific Properties

This operation depends on the total paint system. Necessary specific properties include resistance to all the destructive forces to which the paint system will be exposed.

3.3.3. Paint Guidelines

The following guidelines illustrate the need for the paint-system approach. By considering a paint system before beginning to paint, we can be sure of selecting compatible and appropriate component coats.

3.3.3.1. The Correct Techniques and Tools

Good paint application enables the paint to do the job for which it was formulated and specified.

3.3.3.2. Uniform Wet (and Dry) Film Thickness

Failing to create uniform film thickness can result in a lack of film continuity, leading to serious film defects. The most serious of these defects are

- loss of required resistance properties due to film porosity caused by too dry an application
- loss of resistance properties from poorly cured spots caused by too thick an application, or from thin spots caused by an uneven application
- nonuniform gloss or color, which spoils the decorative function.

3.3.3.3. Correct Number and Sequence of Applications

Manufacturer's directions should provide this information for each paint in the system.

3.3.3.4. Other Guidelines

Some of the most important guidelines to keep in mind are listed here.

- Apply the proper number of wet applications of each paint to give the specified dry-film thickness without trapping the solvent.
- Carefully observe the proper drying or curing time between successive applications of the same paint and between successive paints.
- Follow the manufacturer's directions so that the coating develops the

properties for which it was specified. Failure to follow directions voids the manufacturer's responsibility for the failure of the coating to perform. If the manufacturer's directions are faithfully followed, but the coating does not meet specifications after application, subsequent application should cease at once, and the manufacturer's technical staff should be consulted immediately.

- Remove the film masking after the paint sets but before it is hard to avoid marring.
- Clean painting tools, including brushes, rollers, and spray guns, before the paint has dried or before the catalyzed coating is converted into an insoluble gel.

3.3.3.5. Technique and Tools

The correct tools for applying any paint are those that produce the properties for which the paint was specified and produce them under the actual application conditions. The manufacturer's directions should clearly state the

- proper tools for applying the paint
- proper additions if any to apply the paint correctly
- proper rate of application in square feet per gallon, dry-film thickness, or the equivalent
- any deviations in application technique from the normal use of the application tools, such as spray pressure.

The proper application tool depends on the rheological (flow) properties of the paint being applied. Paints applied by brush should be sufficiently

- thick to lay down an adequate film of wet paint
- nonfluid (as manufactured) to stay on the brush when picked up for an application
- fluid after application to smooth out brush marks
- nonfluid after brush marks have been removed and before the paint sags or drips.

Paint applied by roller must be similar to brushing paint, with the following differences.

- Paint must stay wet longer to cover the roller pattern.
- Paint should be less fluid (as manufactured) to avoid dripping when picked up and applied by the roller.

- Paint applied with a stipple roller must be thicker, more adhesive, and not flow after application.

Paint applied by an air spray gun should be

- nonsticky when thinned to spraying viscosity so that it breaks up easily into droplets
- stable when thinned to a viscosity low enough for spraying
- thinned paint, ready for application, containing a solvent blend volatile enough to evaporate primarily in the spray cone before the paint reaches the surface; the solvent must also keep paint wet long enough to be smoothed out after being deposited on the surface.

Paint atomized by high-pressure pumping (airless spray) rather than being broken up by a large volume of air mixed with it requires paint with the following properties.

- The paint can be sprayed at higher solids (higher percentages of pigment and resin in the formulation) and higher viscosity, with a thicker film applied at each pass of the spray gun.
- The solvent blend need not evaporate so quickly because there is less solvent in the paint at application viscosity.

3.3.3.6. Correct Drying Cycle

The final properties of the dried or cured coating develop during the drying cycle. Unless conditions are correct, the desired film properties will never develop. The coating composition determines the conditions for proper drying or curing, but usually, only manufacturers of coatings used in the paint system know the exact composition of the coatings. However, compositions can be determined through reverse engineering.

3.4. PROTECTING AGAINST WATER AND OTHER FAILURES

Water is the hidden enemy of paint because it is a pervasive element of deterioration. Water goes everywhere, either as a liquid, a vapor, or a carrier for chemicals dissolved in water. Water produces the following effects:

- rusting and other corrosion
- paint peeling

- masonry efflorescence and spalling
- wood rot
- corrosive water solutions (*e.g.*, staining, seawater).

3.4.1. Rusting and Other Corrosion

The corrosion of metals from exposure to water can be prevented if the proper paint system is correctly applied according to the following guidelines.

- Make sure that the surface is properly prepared to receive and hold paint. Remove spots of metal corrosion, such as rust, and all loose material before priming the metal.
- Choose a primer with good water resistance that adheres to the metal.
- Make sure that the total paint system is water resistant and durable under the conditions of actual exposure.
- Coat the entire metal surface exposed to water, not just the part that is easy to reach and visible.

3.4.2. Peeling Paint

This problem usually occurs when water is trapped behind the paint. If the paint system is not sufficiently resistant to water for the intended exposure, water can penetrate the surface. However, peeling is usually caused by water that has entered through breaks in the paint surface due to cracks in the wood or open joints that were not properly caulked. Special care should be taken to caulk joints around doors and windows.

Another cause of paint peeling is excess moisture in the space behind a painted surface. The moisture condenses into water, enters the unpainted portion of a surface, then pushes the paint off the surface. If the excess moisture comes from inside a building, it is usually caused by insufficient ventilation to the outside and a paint system that is not waterproof. The solution in this case is to provide more ventilation to the outside and if necessary to apply a waterproof paint system. If the excess moisture comes from the outside, it is usually caused by insufficient ventilation of the space between the inside and outside surfaces. The solution in this case is to increase ventilation by adding ports to an exterior surface. Excessive caulking and sealing can turn a house into a blister box, resulting in excessive peeling when there are insufficient ventilation ports.

Unpainted or nonwaterproofed areas, such as the lower edges of wood siding and stucco walls, top and bottom ends of uncaulked, unpainted exterior doors, and open joints between siding that are too small to caulk, are a common source of water in painted walls. Open joints should be waterproofed with repeated applications of a penetrating wood preservative until the ends of the siding are saturated. The amount of penetrating wood preservative or wood primer required to protect bare wood is far greater than generally realized. The ideal treatment involves a 24-hour cold soak in a penetrating wood preservative to saturate a wooden surface. Paint should be applied over the treated wood after drying.

When specifying penetrating wood preservatives or paint primers, it is important to understand how greatly the porosity of wood varies with the kind of grain exposed. Face grain, the least porous, requires the smallest amount of penetrating wood preservative or paint primer to satisfy the porosity of the wood. Edge grain, which is about twice as porous, requires about twice as much preservative or primer; the edges of boards are usually edge grain. End grain is about four times as porous as face grain and requires about four times as much preservative or primer; the ends of boards are all end grain. Slash grain is a combination of edge grain and end grain.

Because of the way plywood is constructed, its face is a mixture of face grain, edge grain, and end grain. All sawed edges of plywood contain layers of end grain, and when used outside, sawed plywood should be waterproofed and caulked. Although exterior-grade plywood is made with waterproof glues that will not delaminate, plywood does delaminate if water penetrates through the surface or the edge, because the layers of wood swell and split between the layers of waterproof glue. For this reason, it is important to seal the edges of plywood.

A third common cause of peeling paint is moisture or water from the moist ground under a house or structure. For this reason, the bare ground in the crawl space under the structure is often covered with tar paper. Caulking breaks in floors and foundations may help alleviate the problem, but in severe cases, it may be necessary to remove the source of excess water in the soil.

3.4.3. Masonry Efflorescence and Spalling

Efflorescence refers to the deposit of water-soluble salts on the outside of painted masonry surfaces, such as stucco, concrete, volcanic-ash cinder block, and mortar between blocks. Water-soluble salts are leached out of the masonry when water penetrates the masonry from the inside. When water evaporates from the painted surface, it leaves a deposit of salts. Although the salts can be washed off with a stream of water, they will form again unless the source of water behind the paint is removed. Many times, the source is a leaking roof.

If the efflorescence is only with about 16 in. (406.4 mm) off the ground, it is probably due to the capillary action of the stucco absorbing groundwater. Efflorescence due to capillary action and groundwater will eventually dissolve enough soluble salts from the stucco to cause it or lose integrity and become powdery. Then the stucco will crumble or flake off when touched, taking the paint with it. Capillary action can be stopped by keeping the ground around the base of buildings dry. It can also be stopped by waterproofing the foundation with tar or waterproof paint for 2 ft (0.6 m) below ground level and a few inches above ground. If the building is coated with an emulsion paint that readily transmits water vapor, efflorescence will occur. If the paint (*e.g.*, an oil paint) is impervious to water vapor, it will probably peel off.

Spalling refers to masonry flaking off when the water in it freezes; the paint, of course, comes off with the masonry. Spalling results from the same moisture sources as does efflorescence, and spalling can be corrected in the same ways as efflorescence.

3.4.4. Wood Rot

When water penetrates painted wood, wood rot occurs. The water can come from leaks, inadequately vented space between walls, excessive moisture inside an inadequately vented building, through porous wall paint, or from groundwater absorbed by the wood behind the paint due to the strong capillary action of the wood. Keeping wood at least 6 inches (152.4 mm) above the ground eliminates the source of unwanted water; if the wood cannot be cut, the earth should be dug away at least 6 inches below the wood.

3.4.5. Corrosive Water Solutions

Such water solutions contain dissolved material. Depending on the nature and amount of the material, they usually accelerate the deterioration of some function or property of the paint. For example, dissolved compounds of iron rapidly deteriorate the decorative function of paint as rust stains form on the surface, spoiling its appearance. These stains are difficult or impossible to remove without damaging the paint film.

Seawater contains large amounts of dissolved salts that accelerate water penetration of the paint film causing water to come in contact with the substrate.

3.5. PAINT INGREDIENTS AND FUNCTIONS

Paint components fall into four categories: vehicles, solvents, pigments, and additives. The vehicle portion of paint gives it film continuity and provides adhesion to the substrate. This component is called the vehicle because it carried ingredients to the substrate that will remain on the surface after the paint has dried. The vehicle contains the film former, which is a combination of resins, plasticizers, drying oils, *etc*. In liquid paints, the vehicle includes all the liquids in the paint and all the additives needed by those liquids. Solvents are low-viscosity, volatile liquids used in coatings to improve application properties.

Pigments give the coating properties that cannot be obtained from the vehicle alone, such as

- opacity, color, and gloss control (decorative function)
- specific properties, such as hardness, corrosion resistance, rapid weathering resistance, abrasion resistance, and improved adhesion (protective function)
- desired specialized functions, such as ease of sanding, flame retardance, and electrical conductivity.

Pigments are also used to fill space in paint films. This important function is often abused by adding excessive amounts of filler, inert, or extender pigments to reduce the raw material cost of the paint.

Additives modify the paint vehicle or paint pigments (or both), for example, by improving the drying speed (vehicle), resistance to fading (pigment), or the ease of application.

Paint components, or raw materials, must be carefully controlled to produce coatings of consistent quality. A close inspection of incoming raw materials for conformance to preset quality standards is an important aspect of quality control.

3.5.1. Paint Vehicles

Paint vehicles can be divided into the following six groups according to how they form films and how these films dry or cure.

- Solid thermoplastic film formers. The solid resin is melted for application and solidifies after application.
- Lacquer-type film formers. The vehicle dries by solvent evaporation.
- Oxidizing film formers. Oxygen from the air enters the film and cross-links it to form a solid gel.
- Room-temperature catalyzed film formers. Chemical agents blended into the coating before application cause cross-linking into a solid polymer at room temperature.

- Heat-cured film formers. Heat causes cross-linking in the film former or activates a catalyst.
- Emulsion-type film formers. The solvent evaporates, and suspended droplets of plastic film former flow together to form a film. The plastic droplets are not soluble in the solvent, which is usually water.

3.5.1.1. Solid Thermoplastic Film Formers

An old example of these vehicles is melted tar that resolidifies on cooling. A new application of this type of drying mechanism involves immersing a hot object in powder coatings. Resin in the powder coating melts, causing the coating to adhere to the object, and subsequently reheating the object improves the coating fluidity.

3.5.1.2. Lacquer-Type Film Formers

The curing of a lacquer is usually attributed to the solvent evaporating. Although the drying process is not that simple, the explanation is adequate. The most familiar type of lacquer is based on nitrocellulose, which produces rapid and hard drying. The formula also includes one or more softer resins for adhesion and one or more plasticizers for flexibility. A complex solvent blend is normally used to give a controlled evaporation rate and ensure that all components remain in solution until solvent evaporation is complete.

3.5.1.3. Oxidizing Film Formers

These film formers are based on drying oils that react with oxygen in the air to cross-link the molecules into a solid gel. The cross-links are oxygen bridges between the drying oil molecules. Common drying oils include linseed, soybean, castor, safflower, and tung (china wood) oils, fish oils, and tall oil (from pine trees and a by-product of kraft paper manufacture). Oxidizing film formers can be based on drying oils, varnishes, or oxidizine alkyds.

3.5.1.4. Varnishes

These vehicles are made by cooking drying oils with hard resins. The properties of the varnish depend on the drying oil, the resin, the ratios of these to one another, and processing conditions. Familiar resins used in varnishes include phenolic, ester-gum, maleic, and epoxy. Urethane varnishes are sometimes called urethane oils because of their low viscosity and great flexibility.

Short-oil varnishes contain more resin and less oil, which makes them harder, more brittle, and faster drying. Long-oil varnishes contain more oil and less resin, which makes them softer, more flexible, and slower drying. Medium-oil varnishes are midway between short and long oils in composition and properties.

3.5.1.5. Alkyds

These vehicles are synthetic drying oils in which a larger or smaller part of the drying oil molecule has been replaced by a synthetic molecule, resulting in a new molecule with superior properties. In varnish, resin is dispersed in the oil gel; in an alkyd, a totally new molecule is formed. For this reason, alkyds were called synthetics for many years after their introduction. Air-drying alkyds dry at room temperature because of the amount of drying oil present in the molecule. Like varnishes, alkyds are classified as short, medium, and long oils to describe differences in drying-oil content and the resulting differences in properties. Alkyds made with nondrying oils, such as coconut oil, are used in heat-cured film formers and plasticizers.

3.5.1.6. Room-Temperature-Catalyzed Film Formers

Catalyzed film formation and cross-linking to produce a solid gel differ from the formation of oxidizing film formers. In catalyzed coatings, the oxygen bridge is not an important part of the cross-linking. Instead, chemical agents called catalysts cause a different type of direct chemical bonding. Some catalysts are effective in small amounts that do not constitute a significant portion of the film after curing; other catalysts are used in large amounts and constitute a significant portion of the film after curing. Curing produces a new polymer in which the original reacting molecules are an integral part of the polymer and can no longer be identified. This direct linking without oxygen bridges possesses superior chemical and age resistance. At room temperature, catalyzed coatings produce the benefits of heat-cured film formers.

3.5.1.7. Heat-Cured Film Formers

Heat causes direct cross-linking between molecules in the film former or activates a catalyst that is inactive at room temperature. The superior properties of heat-cured films are similar to those of room-temperature catalyzed coatings. Since it is necessary to heat apply under controlled conditions, these coatings are normally industrial finishes. Results obtained with these finishes are more consistent than those obtained with room-temperature catalyzed coatings because application and curing conditions can be more closely and consistently controlled.

3.5.1.8. Emulsion-Type Film Formers

In its simplest form, an emulsion-type film former consists of drops of plastic floating in water. As the water evaporates, the drops come together to form a film. The spherical drops of plastic assume the form of pancakes and overlap and stick together. Plasticizers are added to make plastic film more flexible and improve adhesion. Coalescing agents that act as (slow-evaporating) solvents for the plastic are added to improve film knitting. The other usual film ingredients, such as pigments, are also present.

3.5.2. Solvents

Solvents (Mellan, 1977) have been defined as low-viscosity volatile liquids used in coatings to improve application properties. Most liquid coatings cannot be applied without them, although many new solventless coatings are now available.

3.5.2.1. Function

Paint solvents simultaneously perform two or all three of the following important operations.

- Dissolve the film former of solution coatings by separating and keeping droplets of a film former apart in emulsion coatings.
- Reduce the solution or emulsion to the proper solid and viscosity for good application properties.
- Control the rate at which paint forms a nonsticky film by their own evaporation rate; this effect extends to the final drying cycle, speeding or slowing it.
- Control the evaporation rate to give good application properties and ensure that the last solvent to evaporate keeps the film former in true solution; the latter is necessary to ensure good film formation and continuity.

True solvents dissolve the film former unaided, while latent solvents dissolve the film former when they are enriched with true solvents. Diluents are tolerated by the film former solution after the former has been dissolved by true and latent solvents. Diluents are normally added to reduce the cost of the mixed solvent, since they act as an extender to the solvent portion of a coating. Generally, diluents must evaporate quickly, before true solvents, otherwise the film-forming resin precipitates from the solution, causing such paint problems as blushing (the milky look) of clear coatings.

3.5.2.2. Types

Coatings solvents are often classified as hydrocarbons and oxygenated solvents; a less important group consists of terpenes. Hydrocarbons contain only hydrogen and carbon atoms in their molecules. The molecules of oxygenated solvents also contain oxygen atoms, but some of the oxygenated solvents in common use contain atoms of other chemical elements, such as nitrogen. Oxygenated solvents are more expensive than the hydrocarbons and usually less readily available. Oxygenated solvents are required when blending solvents for most lacquers and catalyzed coatings and for many synthetic resin solutions. Oxygenated solvents are also present to some extent in practically all the exempt solvent blends required by air pollution control regulations.

There are two types of hydrocarbon solvents: aliphatic and aromatic; primarily aromatic blends of the two types are called semiaromatics. Aliphatic and aromatic hydrocarbons differ in the way the carbon atoms are connected in the molecule. This characteristic structural difference leads to a sharp difference in chemical and toxicological properties. Aromatic hydrocarbons are stronger solvents for coatings film formers; they are also more irritating to human beings in both liquid and vapor forms.

Typical pure aromatic hydrocarbons include benzene, toluene, and xylene; typical pure aliphatic hydrocarbons are hexane, heptane, and odorless mineral spirits. Ordinary mineral spirits are mostly aliphatic hydrocarbons. In the United States, aromatic and aliphatic hydrocarbons are usually derived from petroleum by heat distillation. They are generally used as a mixture of aliphatic and aromatic components available in an extensive range of solvent strengths and evaporation rates.

Oxygenated solvents are manufactured by a variety of processes. Those most commonly used in coatings are known as alcohols, esters, ketones, and glycol ethers. More than one member of each of these chemical types is in common use in coatings. Examples of alcohols include methanol (methyl or wood alcohol), ethanol (ethyl or grain alcohol), isopropanol (isopropyl or rubbing alcohol), and butanol (butyl alcohol). Examples of esters include ethyl acetate, isopropyl acetate, butyl acetate, and butyl cellosolve acetate. Examples of ketones include acetone, methyl ethyl ketone (MEK), methyl isobutyl ketone (MIBK), and cyclohexanone. Examples of glycol ethers include cellosolve (ethylene glycol ethyl ether) and butyl cellosolve (ethylene glycol n-butyl ether).

Terpene solvents are derived from the sap of the pine tree; examples are turpentine, dipentene, and pine oil.

3.5.3. Pigmentation

Paint pigments (Bentley, 1960) are solid grains or particles of uniform and controlled size that are permanently insoluble in the coating vehicle. This insolubility differentiates pigments from dyes: At some stage in their use, dyes are almost always soluble in the carrier in which they are being used. Paint pigments must be nonreactive at all times to work, while dyes must almost always be reactive at some stage to work. Paint pigment properties, like those of paint vehicles, must be controlled to close tolerances to produce paints of consistent quality.

3.5.3.1. Pigmentation and Decoration

In a coating, pigments contribute opacity, color, and gloss control. Desired pigmentation properties are achieved through proper formulation, compounding procedures, and quality control.

Opacity is the ability of paint to hide or obscure the substrate. Opacity is a function of the pigment's index of refraction, which is a numerical measurement of the pigment's ability to bend light rays striking its surface. White, yellow, and orange pigments have the lowest indices of refraction, while black has the highest; "clean" colors have the least opacity. "Dirty" colors, which are formed by the addition of a dark or muddying component, have greater opacity than clean colors. "Dry hiding," sometimes called "high dry hide" or "dry hide hiding" refers to an increase in white opacity caused by using excessive extender or filler pigments in white or tinted paints.

Color results from the ability of pigmentation to absorb certain wavelengths of visible light and reflect other wavelengths. Color is such a complex technical subject that only its general role in paints is discussed here.

Colored pigments differ widely in properties and cost because they are derived by different chemical processes or differ in chemical composition. Colored paints may vary in specific properties even when they appear to the unaided eye to be exactly the same color. The necessary specific properties of a paint for a specified use must include performance characteristics of the color, because it is possible to make a color from many different pigment combinations. Yet pigment combinations can differ widely in resistance properties.

One important resistance property of paint colors is resistance to fading from exposure to light, chemicals, or heat; another is resistance to bleeding— the migration of color from one coat of paint to a succeeding coat of paint.

Bleeding is caused by the solubility of a pigment or dark-colored resin in the solvent or resin of a succeeding coat of paint. The specific resistance properties needed in the paint color for a given use can be determined only by exposing pigments in the color to similar conditions. The manufacturer of the pigment has usually exposed the pigment and compiled the test results. This information is readily available from the manufacturer, who should be able to estimate the durability of any pigment blend requested. For these reasons, the paint user should rely on the advice and integrity of the paint manufacturer when choosing color. This is especially true for an exposure that is liable to cause premature color failure, such as exposure to full sunlight for most of the day, marine exposure, or exposure to corrosive fumes.

Pigments control the gloss of paints by affecting the texture of the coating surface. If a very smooth surface is desired, the formulator should choose pigments that produce only minute roughness. Formulating with larger pigments produces a rougher texture, and in this way the mirrorlike specular reflectance of a clear varnish or a gloss enamel becomes the more subdued diffuse reflectance of a stain varnish or a semigloss enamel. Still higher pigment loadings and/or larger sized pigments produce a flat finish with no gloss at all, sometimes termed "dead flat."

3.5.3.2. Pigmentation Protection

Pigments contribute to the protective function of paint because a properly chosen pigment combination helps achieve the properties required for the specified use. Pigments are normally blended, as in color formulation, and the properties of the blended paint reflect the properties of the individual pigments in the blend. Pigments contributing to the protective function usually appear either in primers or topcoats with protective functions rather than in intermediate coats.

Pigments in primers contribute to substrate adhesion and protection. By distributing the stresses of the drying or curing primer film, pigment particles keep stresses from causing the primer film to shrink excessively. Shrinkage is undesirable because it tends to pull the paint film loose from the substrate. Pigment also provides "tooth," or texture, to which subsequent coatings can cling. Priming pigment also helps distribute the forces of shrinkage caused when the next coating dries. Cumulative shrinkage of successive coats of paint can pull the entire paint system loose from the substrate. By changing the relation of the film former to the substrate, priming pigments help protect the substrate. For example, in metal primers, correct priming pigments increase the long-term adhesion of the primer by a process sometimes called passivation of the metal. Typical metal priming pigments include red lead, zinc chromate, and red iron oxide; these are sometimes blended together and with other pigments in primer formulations.

Pigments in protective topcoats reinforce the properties of the film former. In exterior metalized coatings, aluminum or other metal flakes significantly increase the weathering resistance of the film former by shielding it from ultraviolet degradation and other destructive forces. As a result, many aluminum finishes are formulated with inferior film formers to keep the cost down. The now-restricted lead pigments contribute to the durability of exterior finishes by forming compounds, called soaps, that reinforce the film. Their spiny balls caused gradual erosion of the paint film rather than some other less desirable failure.

Some pigments add to the protective feature of exterior coatings by mechanically reinforcing the film; talc fibers and mica flakes are typical examples. Zinc oxide contributes to film hardness and water resistance due to the zinc soaps formed with drying oils. Zinc oxide and mica also screen out deleterious ultraviolet rays.

3.5.3.3. Pigmentation and Specialization

Ease of sanding is a good example of a specialized paint function that is increased by pigmentation. Using a short-oil varnish with the proper pigmentation further enhances the ease of sanding. Texture, which can reduce slippage, is another specialized function made possible by pigmentation. Fire-retardant paints depend on specialized pigmentation that responds to heat by causing a chemical reaction that reduces the flammability of the coating.

3.5.3.4. Pigmentation as Filler

Most flat paints, such as wall paints and primers, contain a certain amount of pigment to fill space in both the wet- and the dry-paint films. In most lower cost paints, filler pigments are present in sizable amounts to reduce cost, since the cost per unit of volume of filler pigment is much lower than that of either film former or masking pigments.

The vehicle of quality paints, especially wall paints and primers, are sticky if used in high concentration because they are well-polymerized resins. This high degree of polymerization makes the product viscous and also improves adhesion, toughness, and chemical and weathering resistance. In addition, polymerized resins prevent "striking in" due to their resistance to substrate suction. These well-polymerized film formers must be diluted with solvents before application. However, the lower solid level resulting from this reduced viscosity provides too thin a film if applied alone or with a minimum pigment, which uses only prime pigment; that is, pigment necessary for decorative, protective, or specialized functions only.

Up to a certain volume ratio, the addition of selected filler pigment actually reinforces and improves film former properties. Beyond that point, especially for exterior finishes, a larger amount of filler pigment (or prime pigment, for that matter) leads to rapid deterioration of film properties. The volume relationship of pigment to total film-forming solids is called pigment volume concentration (PVC); it is expressed in percentages. The PVC passes through a critical value called the critical pigment volume concentration (CPVC). When that critical volume relationship is exceeded, resistance properties of paint rapidly deteriorate: The film former no longer bridges all the voids between the pigment particles; striking increases.* The film becomes porous, and as a result, it weathers much faster outside; it is less easily cleaned; and it loses some of its abrasion resistance, flexibility, as well as other desirable properties. Proper formulation allows a coating to migrate into a porous surface.

3.5.4. Paint Additives

Paint additives (Rothenberg, 1978), the ingredients added to the film former, pigmentation, and solvents, impart necessary paint properties not supplied by the other ingredients and augment properties not present in sufficient degree. The most familiar example is the addition of driers, which act as catalysts for oxidizing film formers to oil-based paints to speed drying. Paint driers are soaps formed by the reaction of an organic acid with a metal oxide. The most common driers are formed by the reaction of compounds of cobalt, lead, calcium, zinc, zirconium, and other metals with tall oil acids to form the tallates, with naphthenic acid from petroleum to form the naphthenates, *etc.*

Other types of additives include wetting agents, which help disperse pigments during the manufacturing process, and defoamers, which help break up foam generated during agitation and application. Antisettling agents reduce pigmentation settling when paint is stored. Preservatives and fungicides are additives that improve the paints' storage stability by destroying organisms that grow on certain kinds of cured paint films, causing unsightly dark blotches.

Ultraviolet-screening agents that absorb ultraviolet rays from the sun in exposed coatings and reduce ultraviolet degradation in the film, which drastically reduces the life expectancy of certain vehicles, such as the vinyl chloride resins. Plasticizers increase the flexibility and adhesion of film formers; in fact,

---

*Striking occurs when the system specification is incorrect or an inexpensive paint is used.

many of the important film formers, such as the vinyl chloride resins, would not be usable without them. Acid acceptors accept and neutralize acids formed by certain important resins, such as chlorinated rubber, during storage, thereby preventing these acids from causing the film former to deteriorate rapidly. Antiskinning agents reduce or prevent paint skinning in the can, especially during long storage or after some of the paint has been used.

3.5.5. Paint Formulas

A paint formula lists the ingredients: vehicle, solvents, pigmentation, and additives; amounts are normally stated in units of weight for accuracy. Precise metering equipment permits the liquids to be measured in units of volume. The significant relationships among the ingredients of the dried paint film are volume relationships, not weight relationships.

A film former may be present as drying oil, varnish, resin solution, dry resin, plasticizer, or some combination of these. A solvent may be present as free solvent or a component of varnishes or resin solutions. Pigments and additives are usually listed separately.

Differences between the ratios of the principal ingredients (Table 3.1) account for the most important differences between paint types. The most significant ratio is the volume of pigmentation in the dried film compared with the total volume of dried film. Paints with different ingredient ratios include clear finishes, stains, gloss enamels, semigloss (satin) enamels, flat paints, sealers and primers, house paints (for wooden siding), stucco paints, and filling and caulking compounds.

3.5.5.1. Clear Finishes

These materials normally consist of pure film former plus solvent and additives, such as a drier and an antiskinning agent. Clear finishes may be oil and resin cooked into a varnish or a synthetic resin solution, such as an alkyd. Such finishes are normally transparent unless color is added in the form of a pigment or a dye, and they are glossy unless a flatting pigment is added. Properties depend on the oils and resins used and the conditions under which the varnish is processed.

3.5.5.2. Stains

Stains contain a low amount of both film former and pigmentation, especially if they are penetrating stains. They are designed to soak into the surface to give color and some protection without forming a paint film. The so-

Table 3.1. Typical Paint Components

Vehicle	Pigments
Nonvolatile vehicles	Opaque
Solvent based	Translucent
Oils	Transparent
Resins	Special-purpose
Driers	pigments
Additives	
Lacquer vehicles	
Cellulosics	
Resin	
Plasticizers	
Additives	
Water based	
Acrylic	
Polyvinyl acetate	
Styrene-Butadiene	
Other polymers and emulsions	
Selected copolymers	
Additives	
Solvents	
Trade sales/maintenance aliphatic solvents, and in some cases, aromatics	
Chemical/industrial solvents, including in some cases aromatics	
Lacquer solvents, such as ketones, esters, and acetates	

Source: Weismantel (1981).

called heavy-bodied stains or heavily pigmented stains contain more solids than regular stains. Such stains are referred to as shake or shingle paints.

3.6. BASIC MATERIALS IN PAINT AND COATINGS

Paint is a mechanical mixture or dispersion of pigments or powders, at least some of which are normally opaque, in a liquid or medium known as the vehicle. The paint vehicle normally consists of a nonvolatile portion that remains as part of the paint film and a volatile portion that evaporates, thus leaving the film. The dried paint film therefore consists of both pigment and the nonvolatile vehicle. The volatile portion of the vehicle is normally required to ensure proper application properties.

The proportion of pigment to nonvolatile vehicle usually determines the glossiness of the dried film. For example, if this proportion is small (*i.e.*, less than 25% of the total nonvolatile volume), the result is probably a glossy film, since there is more than enough nonvolatile vehicle to cover the pigment completely. As the percentage of pigment volume increases, gloss decreases, so that at a 45% PVC, the paint is probably a semigloss; at a 70% PVC, there is little gloss, so the paint is likely to be flat. Paint and lacquer components are listed in Table 3.1.

There are two general types of coatings: the solvent based, which are reducible by an organic solvent, and the water based, which are reducible by water. The specific properties of a coating depend almost entirely on the specific properties of its pigments and vehicles and their ratio to one another. Many coatings contain little or no pigments; these are the clear coatings, including clear lacquers and varnishes. Clear coatings normally dry to a high gloss, but pigmented clear coatings dry to a dull finish. Special flatting types of pigments that give no color and have no obliterating properties are normally used in these dull-finish clear coatings.

3.6.1. Oils

Oils (Martens, 1974) are used in coatings either by themselves, as a portion of the nonvolatile vehicle, or as an integral part of a varnish when combined with resin, or of a synthetic liquid when combined with its resinous portion. Table 3.2 lists oils used in the coatings industry in 1977. In particular, oil

- improves the flexibility of the paint film; eliminating oil from certain formulations would cause the film to crack
- gives durability to exterior finishes

Table 3.2. Oils Used in the Coatings Industry in 1977

Oil	Millions of Pounds/Year
Linseed	153
Soybean	79
Tung	10
Fish	9
Castor	8
Tall	5
Coconut	4
Other primary oils	7
Total	275

Source: Huber (1978).

- improves gloss
- gives moderate resistance to water, soap, chemicals, and other corrosive products
- provides specialty properties, such as wrinkling for wrinkle finishes
- improves leveling and flow, nonpenetration, and wetting properties of the vehicle when treated.

3.6.1.1. Composition

Most of the oils used in paint are triglycerides of fatty acids. Glycerin, $C_3H_5 (OH)_3$ has three OH groups, each of which can react with the carboxyl group of a fatty acid. Such a reaction results in water being split off and a triglyceride being formed. The composition and properties of oils are listed in Table 3.3.

3.6.1.2. Properties

Properties of the specific oil depend largely on the type of fatty acids in the oil molecule. Thus, highly unsaturated fatty acids give improved drying properties but have a greater tendency toward yellowing. Drying is especially improved if the double bonds are in a conjugate system where two double bonds are separated by a single bond. Such oils also have a faster bodying (increasing molecular weight and viscosity) rate when heated and somewhat better water and chemical resistance. The composition of fatty acids that compose oils are listed in Table 3.4.

3.6.1.3. Oil Treatments

Many of the oils cannot be used in the raw state, since they are produced by crushing seeds, nuts, fish, *etc*. Others can be used in the raw state but are often treated to give them special properties; among these treatments are the following.

- Alkali refining. The oil is treated with alkali, which lowers its acidity and makes it less reactive and also improves its color.
- Kettle bodying. The oil, usually refined, is heated to a high temperature for several hours to polymerize it. This increases its viscosity and improves its color retention, nonpenetration, and flow, gloss, wetting and drying properties; however, the process impairs brushability.
- Blowing. Air or oxygen is passed through the oil at elevated temperatures. The resultant oil has improved wetting, flow, gloss, drying, and

Table 3.3. Fatty Acid Composition of Vegetable and Marine Oils

	Number of Carbons	Double Bonds	Tung	Olitica	Dehydrated Castor	Fish	Linseed	Safflower	Soya	Tall[a] Oil Acids	Cottonseed	Coconut
Oleic	18	1	8	6	9	10	22	13	25	46	24	7
Linoleic	18	2	4		82	15	16	75	51	41	40	2
Linolenic	13	3	3				52	1	9	3		
Eleostearic	18	3	80									
Licanic	18	3		78								
Ricinoleic	18	1			9							
Palmitoleic	16	1										
Arachidonic	20	4				30						
Clupanodinic	22	5				25						
Stearic	18	0	1	5	2	2	4	4	4	3	4	6
Palmitic	16	0	8	7		12	6	6	11	5	29	11
Myristic	14	0				6					1	18
Lauric	12	0										44
Capric	10	0										6
Caprylic	8	0										6
Iodine Value							170–190	140–150	120–140	128–138	99–138	7–10
Viscosity, Gardner–Holdt							A	A	A	A	A	A
Lb/gal							7.76	7.70	7.70	7.53	7.65	7.68
Color, Gardner							11	10	10	4	8	5
Acid Value							4	4	3	194	1	2
Saponification Value							190	190	190	196	190	250

Source: Martens (1974).
Notes: [a]Distilled 1% rosin acids

Table 3.4. Structure of Fatty Acids

Example of Acid	Formula	Nature of Triglyceride Containing Predominant Amounts of the Acid
Stearic	$CH_3(CH_2)_{16}COOH$	Nondrying solid fat
Oleic	$CH_3(CH_2)_7CH=CH(CH_2)_7COOH$	Nondrying oil
Linoleic	$CH_3(CH_2)_4CH=CH-CH_2-CH=CH(CH_2)_7COOH$	Semidrying oil
Linolenic	$CH_3CH_2-CH=CH-CH_2-CH=CH-CH_2-CH=CH(CH_2)_7COOH$	Drying oil
Elaeostearic	$CH_3(CH_2)_3CH=CH-CH=CH-CH=CH(CH_2)_7COOH$	Fast–drying oil
Ricinoleic	$CH_3(CH_2)_5CHCH_2-CH=CH(CH_2)_7COOH$ $\quad\quad\quad\quad\quad \overset{|}{OH}$	Nondrying oil
Licanic	$CH_3(CH_2)_3CH=CH-CH=CH-CH=CH(CH_2)_4CO(CH_2)_2COOH$	Fast–drying oil
Isamic	$CH_2=CH(CH_2)_4C\equiv C-C\equiv C(CH_2)_7COOH$	Drying oil

Sources: Martens (1974), Gooch (1980).

setting properties, but brushability and often color and color retention are impaired. In addition, paints containing blown oils have a greater tendency toward pigment settling.

3.6.1.4. Linseed Oil

This is the largest volume oil used by the coatings industry. It is very durable, yellows in interior finishes, but bleaches in exterior paints, and has good nonsagging properties. Linseed oil provides easy application, good drying, fair water resistance, medium gloss, a medium bodying rate, but poor resistance to acids and alkalies. It is used primarily in house paints, trim paints, and color-in-oil pastes. Alkali-refined and kettle-bodied linseed oil is used in varnishes and interior paints. Linseed oil is an important modifying oil in synthetic alkyds.

3.6.1.5. Soybean Oil

This is a semidrying oil that can be used only with modifying oils and resins, which improve its drying properties. The refined oil has excellent color and color retention. Soybean oil is one of the most important modifying oils in alkyds, and it is used in nonyellowing types of paint.

3.6.1.6. Tung Oil (China Wood Oil)

This oil contains conjugated double bonds and cannot be used in its raw state, since it would dry to a soft, cheesy type of film. In its kettle-bodied state, it gives the best drying and most resistant film of any of the common paint oils. It has a good gloss and good durability; it is used in finishes requiring resistance to moisture: spar varnishes, quick-drying enamels, floor, porch, and deck paints, concrete paints, *etc.*

3.6.1.7. Olticica Oil

This oil's properties resemble those of tung oil but its drying, flexibility, and resistance characteristics are not quite so good. Olticica oil also has somewhat poorer color and color retention; however, it has better gloss and better leveling qualities than tung oil. Olticica oil is normally used as a substitute for tung oil when there is a large price difference between them.

3.6.1.8. Fish Oil

This is a poor-drying oil that cannot be used in its raw state because of its odor. In its kettle-bodied state, it provides relatively easy application and good nonsagging properties; it also has fairly good heat resistance. Fish oil is used in low-cost paints, since it is usually less expensive than other oils.

3.6.1.9. Dehydrated Castor Oil

Raw castor oil is a nondrying oil used in lacquers as a plasticizing agent to make them more flexible. When castor oil is treated chemically to remove water from the molecule, additional double bonds are formed; this makes it a drying oil. The dehydrated oil dries better than linseed oil, although paints made with linseed oil sometimes have a residual tack that is difficult to remove. Dehydrated castor oil offers very good water and alkali resistance—almost as good as tung oil. Castor oil also has excellent color and color retention properties, on a par with soybean oil. Castor oil is used in finishes where color and drying are important—alkyds, varnishes, and quick-drying paints.

3.6.1.10. Safflower Oil

This oil has some of the good properties of both soybean oil and linseed oil. Safflower oil possesses the excellent nonyellowing features of soybean oil, and safflower oil dries almost as well as linseed oil. Safflower oil can therefore be used as a substitute for linseed oil in many white formulations where color retention is important, especially kitchen and bathroom enamels.

3.6.1.11. Tall Oils

Tall oil is not really an oil, but it is often used as an oil or combined with an oil and a resin. Tall oil is a combination of fatty acids and rosin. Normally, it is separated into its different ingredients for use. As a component in alkyds, the fatty acids produce vehicles similar to those made with soybean fatty acids. When limed, tall oil yields a low-cost, high-gloss liquid with poor flexibility that tends to yellow very badly on aging.

3.6.2. Resins

If in coatings, oils were the only nonvolatile components with the exception of driers, the result would be a relatively soft, slow-drying film. Such a film would be satisfactory for house paints, ceiling paints, or other surfaces where hardness and rapid drying are not important but totally unsatisfactory for many trade sales and maintenance coatings and most industrial or chemical coatings. In addition to improving hardness and decreasing drying time, specific resins often improve gloss and gloss retention and adhesion to the substrate. Resistance to chemicals, water alkalies, and acids cannot be achieved without using different types of resins, and low-cost resins reduce the raw material cost of a coating.

3.6.2.1. Rosin

This low-cost natural resin, derived from tree sap, is essentially abietic acid $C_{20}H_{30}O_2$; it must be neutralized before using. This is normally done by reacting the rosin with lime, in which case the product is known as limed rosin; a reaction with glycerin produces ester gum, while a reaction with pentaerythritol yields pentaresin. Liming rosin produces a high-gloss resin with excellent gloss retention and fine adhesion. However, the resin dries slowly and is not resistant to water and chemicals. Since limed rosin tolerates large quantities of water, it is popular in low-cost finishes, and a solution of limed rosin in mineral spirits, called gloss oil, is popular in low-cost floor paints, barn paints, and general utility varnishes.

3.6.2.2. Ester Gum

This resin is made by reacting rosin with glycerol $C_3H_5(OH)_3$, which neutralizes or esterifies the abietic acid. Ester gum can be considered the first synthetic resin. Ester gum dries somewhat more slowly than limed rosin, but it has better color retention and resistance characteristics. Ester gum produces a very high gloss and has excellent adhesion. The higher acid number ester gums are compatible with nitrocellulose and are therefore used in lower cost gloss lacquers.

3.6.2.3. Pentaresin

When pentaerythritol $C(CH_2OH)_4$ is the alcohol used to react with rosin, the result is a resin with a higher melting point, good heat stability, color, color retention, and a high gloss. When the resin is cooked into varnishes with different oils, good drying properties and a moderate degree of water and alkali resistance are obtained. Like other resin esters, pentaresin offers good adhesion to all types of surfaces.

3.6.2.4. Coumarone–Indene (Cumar) Resins

These resins derived from coal tar are essentially high polymers of the complex cyclic and ring compounds of coumarone and indene. They are completely neutral and thus ideal for flake types of aluminum paints. In addition, these resins offer good [alcohol and electrical breakdown] properties. They are also resistant to such corrosive agents as brine, dilute acids, and water. On the negative side, the resins offer poor color retention and only fair drying properties and gloss. Their cost is normally quite low.

3.6.2.5. Pure Phenolic Resins

These pure synthetic resins are made by reacting phenol with formaldehyde. There are two resin types: one that is cooked into oil and used largely in trade sales and marine paints and another that is sold dissolved in a solvent and applied in that form and baked. The first resin type offers excellent water resistance and durability, making it ideal for exteriors, floors, porches, decks, and marine paints or varnishes. Since resin type 1 also offers fine chemical, alkali, and alcohol resistance, it can be used for furniture, bars, patios, and similar applications. In some instances, adhesion is rather poor.

The solvent-type resin is heat reactive, and it becomes extremely hard and resistant to chemicals when properly cured. Resin type 2 is used for lining cans, tank interiors, and related applications. All phenolics tend to yellow.

3.6.2.6. Modified Phenolic Resins

These resins are a combination of ester gum and pure phenolics, and their properties reflect those components. The resins offer very good water, alkali, and chemical resistance, and the ester gum portion give the resins good adhesion. Since the resins offer a good resistance to moisture and a high gloss, they are suited for floors, porches, and decks, sealers, spar varnishes, and any other uses requiring a combination of good resistance, a hard film, and rapid drying and tolerating paint yellowing.

3.6.2.7. Maleic Resins

These resins are made by reacting maleic acid or anhydride with a polyhydric alcohol, such as glycerin, in the presence of rosin or ester gum. They offer very rapid solvent release, good compatibility with nitrocellulose, and good sanding properties. This combination makes the resins ideal for sanding lacquers. Maleic resins also dry rapidly and offer good color retention, so they can be used in quick-drying white coatings. Maleic resins should be used only in shorter oil lengths, since they have some tendency to lose resistance to moisture as they age in longer oil lengths.

3.6.2.8. Alkyd Resins

These resins are manufactured by reacting a polybasic acid, such as phthalic acid or anhydride, with a polyhydric alcohol, such as glycerin and pentaerythritol, then further modified with drying or nondrying oils. These resins are probably the most important ones used in solvent-based trade sales paints and many industrial coatings. Resins modified with large percentages of drying oils are normally used in trade sales paints; they are known as long-oil or medium-oil alkyds. Resins modified with smaller percentages of oil or with nondrying oils are used in industrials, baking finishes, and lacquers; they are known as short-oil or nondrying alkyds. Normally, the larger the percentage of glyceryl phthalate, or resinous portion, the less drying time required, the more brittle the finish, and the better the baking properties. Other properties depend on the type of modifying oil and polybasic acid used.

Generally, alkyds have excellent drying properties combined with good flexibility and excellent durability. Color retention, when modified with non-drying oil or with oil having good retention, such as soybean or safflower oils, is very good. Gloss and gloss retention in alkyd paints are unusually good. In baking finishes, alkyds are normally combined with other resins, such as urea and melamine, to obtain top-grade films. The resistance characteristics of alkyds, though good, do not compare with those of pure phenolics nor equal those of modified phenolics. If high-resistance characteristics are not required, however, alkyds are second to none in good overall properties. Thus, they are ideal for all types of interior, exterior, and marine paints and a large percentage of industrial coatings.

3.6.2.9. Urea Resins

The ureas are obtained from the reaction of urea and formaldehyde. They form a hard, fairly brittle, and colorless film. By combining the ureas with alkyd resins or plasticizers, this brittleness and rather poor adhesion can be corrected. Short-oil, high-phthalic alkyds are combined with ureas and melamines in baking finishes. Urea resins can be used only in baking types of coatings, since they convert from a liquid into a solid form under the influence of heat in a type of polymerization often called curing. The ureas offer excellent color retention and good resistance to alcohol, grease, oils, and many corrosive agents. The ureas make excellent finishes for many metallic surfaces, such as refrigerators, metal furniture, automobiles, and toys.

3.6.2.10. Melamine Resins

These resins are synthesized from melamine, a ring compound, and formaldehyde; they act much as do urea resins. However, melamine resins cure more quickly or at lower temperatures and give a somewhat harder, more durable film with higher gloss and better heat stability. Although melamine resins are more expensive than urea resins, they are preferred for high-quality white finishes because their shorter baking cycle produces a film that is whiter and provides the best color retention.

3.6.2.11. Vinyl Resins

Solvent-based vinyl resins are normally copolymers of polyvinyl chloride and polyvinyl acetate, though they are available as polymers of either one. They are usually available as white powders to be dissolved in strong solvents, such as esters or ketones, but they are also available already dissolved in such solvents. Vinyl resins are plasticized to make an acceptable film. Chloride is very difficult to dissolve but has extreme resistance to chemicals, acids, alkalies, and solvents, while acetate is not so resistant but much more soluble. The more practical copolymer still exhibits exceptional resistance to corrosive agents, chemicals, water, alcohol, acids, and alkalies. Vinyl resins do an exceptionally fine job in coatings for cables, swimming pools, cans, masonry, or any surface requiring very high resistance.

3.6.2.12. Petroleum Resins

These completely neutral, rather low-cost resins are obtained by removing the monomers when cracking gasoline and then polymerizing them. Petroleum resins offer good resistance to water, alkalies, alcohol, and heat. Some have good initial color, but they all tend to yellow on aging. Petroleum resins are very good for aluminum paints, and they make good finishes for bars, concrete, and floors when cooked into tung or oriticica oil.

3.6.2.13. Epoxy Resins

These resins, more correctly called epichlorohydrin bisphenol resins, are chain-structured compounds composed of aromatic groups and glycerol joined by ether linkages. Various modifying agents are used to produce epoxies of different properties, but all such resins generally have excellent durability, hardness, and chemical resistance. They can be employed for high-quality air-drying and baking coatings, and some can even be used with nitrocellulose in lacquers.

3.6.2.14. Polyester Resins

In addition to the alkyd resins, which are polyesters modified with oil, there are other types of polyesters, such as polyester polymers, that offer a light color and good color retention, excellent hardness combined with good flexibility, and very good adhesion to metals. They are useful in many industrial-type coatings where such properties are important.

3.6.2.15. Polystyrene Resins

These resins are derived from the polymerization of styrene. They are available with a variety of melting points that depend on the degree of polymerization, and they are thermoplastic. The higher melting point resins are incompatible with drying oils, but the lower polymers are compatible to some degree. Polystyrene resins offer high electrical resistance, good film strength, high resistance to moisture, and good flexibility when combined with oils or plasticizers. These resins are useful in insulating varnishes, waterproofing paper, and similar applications.

3.6.2.16. Acrylic Resins

These thermoplastic resins are obtained by the polymerization or copolymerization of acrylic and methacrylic esters. Acrylic resins may be combined with melamine, epoxy, alkyd, acrylamide, *etc.*, to produce systems that bake to a film with excellent resistance to water, acids, alkalies, chemicals, and other corrosives. Acrylic resins are used in coatings for all types of appliances, cans, and automotive parts and for all types of metals.

3.6.2.17. Silicone Resins

These polymerized resins of organic polysiloxanes combine excellent chemical-resistant properties with high-heat resistance. They are expensive and therefore not usually used for their chemical-resistant properties, which can be obtained from lower priced resins. Silicone resins are used for their very important heat- and electrical-resistant properties, which are superior to those of other resins. Silicone resins can be copolymerized with alkyds at a lower cost and still retain some of their important properties.

3.6.2.18. Rubber-Based Resins

These resins, based on synthetic rubber, provide a film when properly plasticized that has high resistance to water, chemicals, and alkalies. Rubber-based resins are excellent as swimming pool paints, concrete floor finishes, exterior stucco and asbestos shingle paints, and other coatings requiring a high degree of flexibility and resistance to corrosion.

3.6.2.19. Chlorinated Resins

Paraffin can be chlorinated at any level from 42%, which gives a liquid resin, to 70%, which gives a solid resin. Chlorinated resins are popularly used in fire-retardant paints. The 70% resin is also used in house paints and synthetic nonyellowing enamels for improved color and gloss retention. Chlorinated biphenyls with high-resistance characteristics can also be made; these are often combined with rubber-based resins for coatings requiring a high degree of alkali resistance. Rubber is also chlorinated and sold as a white granular powder containing about 67% chlorine. It is quite compatible with alkyds, oils, and other resins, such as phenolics or cumars. Rubber offers high resistance to acids, alkalies, and chemicals, and it is useful for alkaline surfaces, such as concrete, stucco, plaster, and swimming pools.

3.6.2.20. Urethanes

Three general classes of urethane resins or vehicles are available today: amine catalyzed, two-container systems, moisture-cured urethane, and urethane oils and alkyds. The first and second types contain unreacted isocyanate groups that are used to achieve final cure in the coating. In the first case, an amine is used to catalyze a cross-linking reaction, which results in a hard, insoluble film; in the second, moisture in the air acts as a cross-linking agent.

Urethane oils and alkyds, on the other hand, are cured by oxidation in the same way as alkyds and oils, and the urethanes require driers or drying catalysts. However, cure occurs more quickly, and the resultant film is very hard and abrasion resistant, with greatly improved resistance to water and alkalies; color retention is somewhat poorer. Because of their advantages, urethane oils and alkyds are widely used in premium floor finishes and for exterior clear finishes on wood. The hardness of the film tends to impair intercoat adhesion, and care must be exercised to sand the surface lightly between coats to provide tooth.

3.6.3. Lacquers

Lacquers dry when their solvents evaporate. Raw materials consist of substances that form a dry film or can become part of a dry film without oxidation or polymerization and the solvents in which these film formers are dissolved.

The basic film formers of lacquers are then cellulosics; most lacquers also contain resin for improved adhesion, film thickness, and gloss, and plasticizers for improved flexibility. Each of these three types of lacquer film formers is briefly examined.

3.6.4. Cellulosics

The most important cellulosic is nitrocellulose, ethyl cellulose is next in importance, and cellulose acetate is of some importance. Nitrocellulose is made by nitrating cotton linters; it comes in two grades: regular soluble types (RS) and spirit- or alcohol-soluble types (SS). Both are available in a variety of viscosities and form a film that is hard, tough, clear, and almost colorless.

Ethyl cellulose is made by reacting alkali cellulose with ethyl chloride. It comes in different viscosities, and compared to nitrocellulose, it has greater compatibility with waxes, better flexibility, better chemical resistance, less flammability, and a higher dielectric constant. Ethyl cellulose is also somewhat softer than nitrocellulose, tends to become brittle when exposed to sunlight and heat, and is more expensive. These disadvantages can be partially overcome by using proper modifying agents and solvents.

Cellulose acetate lacquers are tough and stable in the presence of light and heat. They also offer good resistance to oils and greases, and they are durable. However, these lacquers have poor solubility and compatibility, and this defect partially limits their usefulness.

In most instances, a lacquer film contains a larger percentage of resin than the cellulosic because resins add many important properties to lacquer films and usually cost less. The most valuable property that resins add is adhesion, which is of particular importance because nitrocellulose by itself offers rather poor adhesion. In addition, resins produce higher solids and therefore a thicker film, improve gloss, reduce shrinkage, and improve heat-sealing properties. When choosing a resin, it must be compatible with the cellulosic being used as well as soluble in a mixture of esters, alcohols, and hydrocarbons to give clear, transparent film.

Common resins in use are rosin esters, such as ester gum, which is used for its low cost; maleic resin, used in wood finishes for its good sanding properties; and alkyds, employed for their good resistance and durability. Alkyds modified with coconut oil are often used; these may be further modified with other resins, such as terpenes, for good heat-sealing properties; and phenolics, for good water resistance.

3.6.5. Plasticizers

Without plasticizers, most lacquers would be much too brittle and tend to crack thereby impairing their durability. In addition to providing flexibility, plasticizers increase the solids content to produce films of practical thickness, and they also tend to improve gloss, especially of pigmented lacquers. Another advantage, especially of chemical plasticizers, is their ability to act as a solvent for the cellulosic and thus enable more of this cellulosic to be used. Plasticizers also help decrease the lacquer's settling time, thus enabling it to flow to form a satisfactorily smooth film.

Plasticizers must be completely nonvolatile to remain in the film permanently. There are some exceptions to this requirement in such lacquers as nail polish, which does not remain on the surface permanently. Since most plasticizers are less expensive on a solids basis than cellulosics, there may be a tendency to use excessive amounts, which would result in a tacky, soft film with poor chemical, water, and abrasion resistance.

Oil and chemical plasticizers are generally used in lacquers. A good example of the nonsolvent oil type is raw and blown castor oil, which provides perpetual flexibility, good color and color retention, is inexpensive, and sensitive to temperature change. Excessive amounts tend however to be spewed from the film. Solvent-type chemical plasticizers, such as dibutyl, phthalate, triphenyl phosphate, and dioctyl phthalate, offer excellent compatibility and good heat-sealing properties. Chlorinated polyphenyls possess good resistance characteristics. All solvent types tend to produce a good, tight film.

3.6.6. Water-Based Polymers and Emulsions

Production of these types of coatings (Stevens et al., 1980) is the fastest growing segment of the coatings industry. Most of the trade sales and architectural paints are not water based. These coatings are water thinnable, and in the case of trade sales paints, they offer little odor, rapid drying, less penetrating, very good alkali resistance, excellent stain resistance, and an easy cleanup with water. In all cases, such coatings release almost no solvent fumes into the atmosphere, which is a considerable advantage given environmental restrictions.

3.6.6.1. Styrene–Butadiene

This is the oldest and initially the only polymer available for latex paints. It is a copolymer of polystyrene, a hard, colorless resin; and butadiene, a soft, tacky, rubberlike polymer. Paints based on styrene edition polymers have some disadvantages in their tendency toward poor freeze–thaw stability and low critical PVC. There is also a greater tendency toward efflorescence, the appearance of a white crystalline deposit on a painted surface. Styrene–butadiene polymers are rarely used now.

3.6.6.2. Polyvinyl Acetate (PVA)

This is one of the most popular polymers used in manufacturing latex paints. The polymer itself is a thermoplastic, hard, resinous, colorless product with good water resistance. Normally, it is obtained as a water emulsion containing surface-active agents, protective colloids, and a catalyst. Polyvinyl acetate is much more stable and easier to use than styrene–butadiene, and has therefore largely replaced it in latex paints. The PVA film is clear, colorless, and odorless, with very good water and alkali resistance. The polymer gives a breathing type of film, which prevents blisters if applied over somewhat moist surfaces. Since by itself the film would be too brittle, it must be plasticized, either internally or in the paint formulation. Polyvinyl acetate polymers are superior to styrene–butadiene polymers in durability, stability to light aging, and nonblistering properties. The emulsion tends to be acidic, and formulating with it requires some caution.

3.6.6.3. Acrylics

Acrylic polymers are probably the best quality emulsions regularly used in manufacturing latex paints. They are essentially made by polymerizing or copolymerizing acrylic acid, methacrylic acid, acrylonitrile and then esterifying them. The properties of acrylic polymers depend to a large degree on the type of alcohol from which the esters are prepared. Normally, alcohols of lower molecular weight produce harder polymers, and acrylates are generally softer than methacrylates.

Unlike PVAs, acrylics are basic (*i.e.*, nonacid), thus reducing the danger that they will cause containers to rust. Moreover, since acrylics are almost completely polymerized prior to application as a paint film, there is practically no brittleness nor yellowing with age. These features improve the durability of acrylic paints; in fact, durability is a special feature of acrylics.

They are the most stable of the polymers and require a minimum of such stabilizers as protective colloids, dispersing agents, and thickeners. Acrylics also withstand extreme temperatures to a high degree. Acrylics have excellent resistance to both scrubbing and wet abrasion; moreover, the extreme insolubility of the dried paint film gives it excellent resistance to oil and grease. As a result, oil stains and other marks are easily removed without injuring the film.

The major disadvantage of acrylics is cost, which is higher than other latices. In partial compensation, acrylics take higher pigmentation, so more low-cost extenders can be used.

3.6.7. Other Polymers and Emulsions

While most trade sales paint is water based, satisfactory water-based polymers with the required properties have not yet been developed for industrial coatings. Nevertheless, much progress has been made, and satisfactory water-reducible coatings have been made for many industrial applications.

3.6.7.1. Water Reducible Resins

The most popular general type of aqueous industrial vehicles is the so-called water-soluble resin. The basic approach involves preparing the resin at a relatively high acid number and then neutralizing it with an amine, such as ammonia or dimethylaminoethanol. A wide variety of resins, including alkyds, maleinized oils, epoxy esters, oil-free polyesters, and acrylics, is produced in this manner. These resins may be either air-dried or baked vehicles. Such driers as cobalt, manganese, calcium, or zirconium can be added as cross-linkers to the baking vehicles. Coatings made with these vehicles are competitive with solvent-based industrials in terms of gloss, film properties, and overall resistance. There is a problem with air-drying efficiency on aging due to the driers chelating with the amines used.

3.6.7.2. Emulsion Vehicles

Emulsion vehicles, particularly acrylic and styrene–acrylic types, are also being promoted for baking industrial finishes. These cure by cross-linking mechanisms, generally through the use of melamine or urea resins. It is more difficult to obtain high gloss with emulsions as compared with water-soluble resins, but because of their higher molecular weight, emulsions can offer advantages in film strength and resistance properties.

3.6.7.3. Copolymers

Some types of polymers, such as acrylates, methacrylates, and acrylonitriles, can be blended or copolymerized to obtain special properties.

3.6.8. Driers

The basic difference between lacquer and solvent-based paint is that lacquer dries by evaporating the solvent, while paint relies on a combination of oxidation and polymerization. Dryers are used to hasten the drying process. Without these additives, paint would dry in days instead of hours, and in many cases, the film would be softer and have poorer resistance properties.

Most driers are organometallic compounds (*e.g.*, resinates, linoleates, and naphthenates) that act as polymerization or oxidation agents, or both. The soaps must be in a form that is soluble in the vehicle. Everything being equal, the more soluble the soaps, the more effective they are as driers. Tall oil driers, based on tall oil fatty acids, are somewhat less soluble than naphthenates based on naphthenic acid. Synthetic acid driers based on octoic, neodecanoic, and similar acids are now the most popular. In addition, the metal portion of the more active driers is normally oxidizable. One theory is that these driers, especially the oxidation catalysts, act in their reduced form by taking oxygen from the air, then become oxidized, pass the oxygen on to the oil or other oxidizable molecule, are reduced again and therefore in a position to take on additional oxygen to pass on to the oxidizable vehicle. This process is repeated until the film is completely oxidized.

3.6.8.1. Cobalt

The cobalt drier containing 6 to 12% metal cobalt is the most powerful drier used by the coatings industry. It acts as an oxidation catalyst, and it is known as a top drier, drying the top of the film. Excessive amounts of cobalt drier will create stresses and strains in the paint film that can result in wrinkling. Though purple in color, cobalt has low tinting strength and will not discolor a paint.

3.6.8.2. Lead

This drier is normally available in strengths containing 24 or 36% metal lead. It is very light in color and thus does not discolor a paint. Lead is a polymerization catalyst and therefore makes an ideal combination with cobalt, since it tends to harden or dry the bottom of the film. Because of lead laws, this type of drier is gradually being replaced by calcium, zirconium, or both. Some lead driers are lead resinates, lead linoleates, and lead naphthenate.

3.6.8.3. Manganese

This drier, available normally in strengths of 6, 9, or 12% metal, is known as a through drier because it acts on both the top and the bottom of the film. However, it is primarily an oxidation rather than a polymerization catalyst and can therefore cause wrinkling if employed in excessive amounts. Manganese is often combined with cobalt and lead to cut the cobalt content and reduce skinning. At other times, manganese is used with lead as a manganese–lead drier combination. Manganese is brownish and tends to discolor paints if used in large amounts.

3.6.8.4. Calcium

This very light-colored drier, which has no tendency to discolor paints, acts as a polymerization agent similar to lead. Calcium also tends to improve the solubility of lead if used in combination with it, and thus calcium makes lead more effective as drier. Calcium is sold in metal contents of 4, 5, and 6%. It is becoming increasingly popular as a substitute for lead in lead-free paints.

3.6.8.5. Zirconium

Like calcium, zirconium is light colored and usually acts as a polymerization catalyst. In lead-free paints, zirconium is often used with cobalt or combined with cobalt and calcium. Zirconium is sold in concentrations of 6, 12, and 18% metal.

3.6.8.6. Other Metals

Other metals are sometimes used as driers; the most popular are iron, useful in colored baking finishes, and zinc, useful as a wetting and hardening agent. Zinc is also used to reduce skinning tendencies in a paint. Cerium, too, is sometimes used as a drier.

3.6.8.7. Nonmetallic Driers

The elimination of lead has focused attention on nonmetallic driers. The most popular of these is orthophenathroline, which often gives excellent drying properties, sometimes superior to those of standard combinations, when used with manganese or cobalt.

3.6.9. Paint Additives

This group of raw materials is used in relatively small amounts to give coatings certain necessary properties (driers really belong in this category). Since additive compositions are not normally revealed by manufacturers, the following discussion refers to trade names. On occasion, additives are used at the job site if problems arise. In such cases, there should be close coordination with and supervision by the paint manufacturer to avoid even greater problems.

3.6.9.1. Antisettling Agents

This group of agents prevents the pigment from settling or separating in the vehicle. Most commonly, this is accomplished by using additives that create a gel structure in the vehicle that traps the pigment so that it cannot settle to the bottom.

3.6.9.2. Antiskinning Agents

These are essentially volatile antioxidants that prevent paint from oxidizing, drying, or skinning while in the can. These agents then volatilize and leave the paint film, allowing it to dry properly once the paint has been applied. The most common antiskinning agents are methyl ethyl ketoximine, which are very effective in alkyds; and butyaldoxine, which are effective in oleoresinous liquids. Phenolics are sometimes used, but they can delay drying time.

3.6.9.3. Bodying and Puffing Agents

These products increase paint viscosity. Without them, paint is often too thin to be sold or used. In solvent-based paints, gelling or thixotropic agents, such as those discussed in the section entitled "Antisettling Agents," may be used, as are liquid bodying agents based largely on overpolymerized oils. In water-based paints, the most common bodying agents are methyl cellulose, hydroxethyl cellulose, the acrylates, and the bentonites. These agents also tend to improve emulsion stability.

3.6.9.4. Antifloating Agents

Most colors used in the paint industry are a blend; that is, black paint is added to white paint to form gray. Therefore, it is important that one color does not separate from the other, which is the function of antifloating agents. Silicones are sometimes used, but they pose serious bubbling and recoatability problems. Special antifloating agents are sold under various trade names.

3.6.9.5. Loss of Drying Inhibitors

Certain colors, such as blacks, organic reds, and even titanium dioxide, tend to inactivate the drier, so the paint loses its drying ability on aging. Agents are therefore introduced to react slowly with the vehicle and to add drier to replace what was lost. In the past, most of these agents have been lead compounds, such as litharge, but these are now being replaced by cobalt-based agents.

3.6.9.6. Leveling Agents

Sometimes a paint does not flow properly and shows brush or roller marks. These can often be corrected by adding special wetting agents that cause the vehicle to set the pigment better.

3.6.9.7. Foaming

This is much more a problem in water-based than solvent-based paints. Bubbles not only make for an unsightly paint when applied, but they result in a partially filled paint can.

3.6.9.8. Grinding Pigments

Unless pigment is properly ground, the result is a coarse film of poor opacity; in a gloss-finish type paint, the gloss is also poor. Certain types of wetting agents tend to improve the ability of the disperser or mill to separate pigment particles more easily and thus obtain a better grind.

3.6.9.9. Preservatives

Almost every formulation based on water must have a preservative for can stability. Until recently, most preservatives have been mercurials, but these are being partially replaced by complex organics.

3.6.9.10. Mildewcides

Most exterior paints experience discoloration due to the growth of fungi or mildew. Until now, this condition has been prevented by including a mercurial in the paint, often in combination with zinc oxide; at present, nonmercurials are also available.

3.6.9.11. Antisagging Agents

When applied, a paint sometimes flows excessively, causing what is known as curtains, runs, or sags. Most bodying or antisettling agents prevent this tendency, and some prevent sag without increasing paint body.

3.6.9.12. Glossing Agents

If the gloss of a solvent-thinned gloss-type formulation is low, it can usually be increased by changing vehicles or pigmentation or increasing the ratio of nonvolatile vehicle to pigment. However, it may be simpler to use an additive.

3.6.9.13. Flatting Agents

Flatness is easy to obtain in regular flat paints, but in clear coatings, such as flat varnishes or lacquers, this quality can be obtained by using special flatting agents, such as amorphous silica.

3.6.9.14. Penetration

Paint penetration is important in stains and paint applied to a poor surface. Most paints, however, require nonpenetration for improved sealing properties and good color and sheen uniformity. This goal is accomplished mainly by agents that establish a gel structure in the paint.

3.6.9.15. Wetting Agents for Water-Based Paint

Many different types of wetting agents are necessary in water-based paints. Some improve pigment dispersions, while others improve adhesion to a poor surface, such as a slick surface.

3.6.9.16. Freeze–Thaw Stabilizers

These are necessary in water-based paints to prevent coagulating or flocculating when the paints are subjected to freezing temperatures. Stabilizers, such as ethylene or propylene glycol, are added to lower the temperature at which paint freezes. An additive that improves the stability of the emulsion can also be used.

3.6.9.17. Coalescing Agents

These agents soften and partially solvate latex particles in water-based paints to help them flow together and form a more nearly continuous film, particularly at low temperatures. Ether alcohols, such as butyl cellosolve and butyl carbitol, are typically used as coalescing agents.

3.6.10. Solvents

There are essentially three types of volatile solvents: a true solvent, which tends to dissolve the basic film former; a latent solvent, which acts as though it were a true solvent when used with a true solvent; and a diluent, a nonsolvent that is tolerated by the coating. Thus, in a lacquer, ethyl acetate is the true solvent, ethyl alcohol is the latent solvent, and petroleum hydrocarbon is the diluent. In a latex paint, water may be considered a true solvent, but in an alkyd enamel, it would be a diluent.

To apply paint, some materials must be used which do not become part of the paint film. With the exception of the newer 100% solids coatings, such as powder coatings, paint simply could not be applied without a solvent, since in most instances the result would be a semisolid mass. Therefore, a solvent's most important task is to reduce viscosity sufficiently so that the coating can be applied by brush, roller, dipping, or spraying. A solvent has other important functions: It controls the setting time of the paint film, which, in turn, controls the ability of one panel of paint to blend with a subsequent panel of paint. In addition, a solvent controls such properties as leveling or flow, gloss, drying time, durability, sagging tendencies, and other characteristics of the wet paint or paint film.

3.6.10.1. Petroleum Solvents

These constitute by far the most popular group of solvents used in the coatings industry. They consist of a blend of hydrocarbons obtained by distilling and refining crude petroleum oil. The most rapid evaporating types, which are obtained first, are used as diluents in lacquers or as solvents in special industrials. Solvents of the intermediate group are used in trade sales paints. The slowest evaporating group, beginning with kerosine and continuing into fuel oils, are used for heating, lubrication, and other applications.

The most important group used in trade sales paints and varnishes consists of mineral spirits and heavy mineral spirits. Mineral spirits are petroleum solvents with a distillation range of 300–400°F (149–204°C). They are sometimes considered a turpentine substitute because the distillation ranges are approximately the same. Because of their low cost, proper solvency, and correct evaporation rate, mineral spirits are probably the most popular solvents in the coatings industry. Normally, they are the sole solvents in all interior and exterior paints with the exception of flat finishes. Special grades of mineral spirits that pass antipollution regulations are now available. Heavy mineral spirits are a slower evaporating petroleum hydrocarbon and an ideal solvent for flat-type finishes. During cold winter weather, a formulator may use a combination of regular and heavy mineral spirits.

The EPA has established new guidelines, based on regulations already adopted in California, that severely limit the amount of solvent in architectural coatings. The recommended limit is 250 g of volatile organic material per liter of paint. This limit also affects water-based paints containing organic free–thaw agents and additives. Architects switching to new high-solids coatings should work closely with the manufacturer to assure proper performance, and they should make certain that application personnel are properly trained to handle the more complex systems.

A faster evaporating petroleum solvent with a distillation range of 200–300°F (93–149°C) known as VM&P naphtha is sometimes used by painters as an all-purpose thinner; however, its rapid evaporation rate may cause the paint to set too quickly. This product is also used by some manufacturers in traffic paints, where a rapid setting and drying time is desirable.

In some industrials and lacquers, a still faster evaporating type with a distillation range of 200–270°F (93–132°C) is desired. In many coatings, this type provides satisfactory spraying and dipping properties. An even faster evaporating type with a distillation range of 130–200°F (54–93°C) is sometimes used when very rapid evaporation and drying are desired, but it may cause blushing or flatting in the paint or lacquer film.

Because of air pollution regulations, the hydrocarbon solvents that had been the backbone of the coatings industry are being phased out and replaced by mixtures that pass the stringent regulations of such states as California, Illinois, and New York.

3.6.10.2. Aromatic Solvents

This group of cyclic hydrocarbons is normally obtained from coal tar or special petroleum fractions distillation. These hydrocarbons are almost pure chemical compounds and much stronger solvents than petroleum hydrocarbons. With the exception of high-flash naphtha, they are rarely used in trade sales coatings; instead, they are used in industrial and chemical coatings where vehicles having weak solvent requirements are not normally used. Since aromatic solvents are pure chemicals, they have regular boiling points rather than distillation ranges. Naturally, those with the lowest boiling points evaporate most quickly and thus dry most rapidly. The most popular of these aromatic solvents are listed here.

- Benzene C_6H_6; boiling point 175°F (79°C); quite toxic; used in paint and varnish removers. It can cause blushing or whitening in a clear film.
- Toluene $C_6H_5(CH_3)$; boiling point 230°F (110°C); very popular in fast-drying industrials and lacquers.
- Xylene $C_6H_4(CH_3)_2$; boiling point 280°F (138°C); popular in industrials and lacquers where slower evaporation is acceptable.
- High-flash naphtha, a blend of slower evaporating aromatics; distillation range is 300–350°F (149–177°C) for brushing-type industrials and lacquers.

These products are also slowly being replaced by others that can pass stringent air pollution requirements.

3.6.10.3. Alcohols, Esters, and Ketones

A great many of these types of solvents are used in industrials and especially lacquers. The following are the more popular solvents of this type.

- Acetone CH_3COCH_3; very strong and very rapidly evaporating; it can cause blushing; used in paint and varnish removers.
- Ethyl acetate $CH_3COOC_2H_5$; a standard rapidly evaporating solvent for lacquers; relatively inexpensive.
- Butylacetate $CH_3COOC_4H_9$; a very good medium-boiling solvent for lacquers; good blush resistance.
- Ethyl alcohol C_2H_5OH; used only in a denatured form; a good latent solvent for lacquers and shellacs; relatively inexpensive.
- Butyl alcohol C_4H_9OH; a medium-boiling popular latent solvent for lacquers.

Other popular ketones used in lacquers are methyl ethyl ketone and the slower evaporating methyl isobutyl ketone. They are very strong and relatively inexpensive. Very slowly evaporating solvents are sometimes used in lacquers to prevent blushing or for brushing application. Among popular products are the lactates, cellosolve, and carbitol.

3.6.11. Pigments

All the raw materials discussed thus far form portions of the vehicle. In nonpigmented clear coatings, these raw materials are all that are used. In pigmented coatings, or paints, it is necessary to add a pigment or pigments to obtain the important properties that differentiate paint from clear coatings. Paints may contain both a hiding, or obliterating, type of pigment and a nonhiding or as it is sometimes known, an extender type of pigment.

One of the most important properties of pigments is to obliterate the surface being painted. This property is often known as hiding power, coverage, or opacity. We frequently hear such terms as one-coat hiding power, which simply means that one coat of paint, normally applied, will completely cover the substrate or surface being painted. However, if a radical change in color is made, two or even three coats of paint may be required, especially if the paint lacks good hiding power.

Another important reason for using pigments is to add to the surface being painted. Pigments are also used because they protect a painted surface. Everyone will recognize red lead as a pigment used to protect steel from rusting. Not so well known are zinc chromate, zinc dust, and lead suboxide.

Other pigments are used to give paint special properties; for example, cuprous oxide and tributyl tin oxide are used in ship bottom paints to kill barnacles, and antimony oxide is used to make paints fire retardant. Pigments can give paint the desired degree of gloss, since everything being equal, the higher the pigmentation, the lower the gloss. Pigments can also give a coating the desired viscosity; control the degree of flow or leveling; improve brushability by accommodating the addition of an easy-brushing solvent; and provide fluorescence, phosphorescence, electrical conductance, or insulation.

3.6.11.1. White Hiding Pigments

White is important not only as a color, but because it forms the basis for a great many shades and tints. The number of important white pigments being used by the paint industry has been dwindling (see Table 3.5). Thus such pigments as lithopone, basic lead, sulfate, titanium–barium pigment, titanium–calcium pigment, zinc sulfide, and many leaded zinc oxides have practically disappeared. Titanium dioxide is the most important white pigment now being used.

Lead carbonate and other lead pigments are not only useful pigments because they impart color, but they are effective mildewcides, so that incorporating lead pigments into a paint formulation usually prevents the growth of microorganisms.

Titanium dioxide (TiO_2) comes in two crystalline forms. The older anatase form has about 75% of the opacity, or hiding power, of the present rutile form, and anatase is less chalk-resistant. Both forms are excellent for interior and exterior use. Titanium dioxide is used in trade sales and chemical coatings. Very little anatase is now being used except in some specialty coatings. Rutile can be used in enamels and flats and solvent-based and water-based coatings;

Table 3.5. Relative Opacity or
Hiding Power of White Pigments

Pigment	Relative Opacity
Basic lead carbonates	15
Basic lead sulfate	15
Zinc oxide	20
Rutile titanium dioxide	125–35

Source: Weismantel (1981).

2–3 lb/gal (240 to 359 kg/m^3) of rutile titanium dioxide provides adequate coverage in most formulations.

Zinc oxide (ZnO) still maintains its importance in the coatings industry, despite its rather poor hiding power (it is only about 15% TiO$_2$) due to unusually good properties that more than offset the relatively high cost of the pigment per unit of hiding power. Zinc oxide's most important use is in exterior finishes, where it tends to reduce chalking and the growth of mildew in house paints. In enamels, zinc oxide tends to improve the film's color retention on aging, and it is sometimes used to improve the hardness of a film.

Extender pigments have practically no hiding power, but they are combined in large quantities with both white and colored hiding pigments. Some extender pigments are used to lower the raw material cost (RMC) of the paint. For example, if prime, or hiding, pigments were used to lower paint gloss for a flat finish, the RMC would be extremely high in most instances; however, extender pigments can achieve this effect at a small fraction of the cost.

Whiting (*calcium carbonate*) is probably the most important extender pigment in use. It comes in a variety of particle sizes and surface treatments, and it can be dry ground, water ground, or chemically precipitated. Normally quite low in cost, whiting can be used to control such properties as sheen, nonpenetration, degree of flow, degree of flatting, tint retention, and RMC.

Talc (*magnesium silicate*) is widely used as an extender in interior finishes, but its greatest use lies in exterior solvent-based coatings, especially house paints, due largely to a combination of durability and low cost. Most talc grades tend to have good nonsettling properties and give a rather low sheen.

China clay (*aluminum silicate*) is used as an extender to some degree in solvent-based coatings, but its greatest use lies in water-based paints. China clay disperses readily with high-speed dispersers in the normal method of manufacturing latex finishes and does not impair paints' flow characteristics. Some grades improve the dry hiding power of water-thinnable or solvent-based paint.

Other extenders sometimes used include diatomaceous silica for reducing sheen and gloss; regular silica, which gives a rough surface; barites, for minimizing the extender effects; and mica, whose platelike structure prevents colors from bleeding.

3.6.11.2. Black Pigments

Next to white, black is probably the most important color in the coatings industry due to its popularity as a color tint, particularly for all shades of gray, which are made by adding black to white.

The two most popular blacks used consist of finely divided forms of carbon known as carbon black and lampblack. Carbon black, the most widely used of the blacks, is sometimes called furnace black. It is made by the incomplete combustion of oil injected into the combustion zone of a furnace. Lampblack, or channel black, is made by the impingement of gas on the channel irons of burner houses. Both types of black come in a variety of pigment sizes and jetness and offer tremendous opacity: Only 2–4 oz/gal (15–30 kg/m^3) of paint are necessary in most instances for proper coverage. Even the most expensive, darkest jet blacks are inexpensive to use because only a small amount is needed. They also have excellent durability, resistance to all types of chemicals, and lightfastness.

While carbon black is used principally as a straight color, lampblack is used mainly as a tinting color for grays, olive shades, *etc*. Largely because of its coarseness, lampblack has little tendency to separate from the TiO$_2$ or other pigments with which it is used and to float up to the surface, as do the carbon blacks with their much finer particle size. Floating, which involves a partial color float to the surface of the film, and flooding, a more nearly complete and uniform color float, are of course undesirable, and for this reason, carbon black is rarely used as a tinting color. Lampblack has very poor jetness but produces a nice bluish shade of gray. It also offers excellent heat and chemical resistance.

Other blacks sometimes used include black iron oxide, as a tinting black with brown tones and in primers, and mineral and thermal blacks, as low-cost black extenders.

3.6.11.3. Red Pigments

There are a great variety of red shades, some of which are discussed in the following paragraphs.

Red tone oxides are good representatives of a series of metallic oxides with very important properties. Though relatively low in cost, they have such find opacity that 2 lb/gal (240 kg/m^3) are normally adequate. In addition, they possess high tinting strength and offer good chemical resistance and colorfastness. These oxides disperse easily in both water and oil, so that high-speed dispersers can be used in manufacturing paints based on iron oxide pigments.

Red iron oxides give a series of rather dull colors with excellent heat resistance. These colors are popular in floor paints, marine paints, barn paints, metal primers, and tinting.

Tolidine reds are popular, very bright azo pigments ranging from light to deep red. They have excellent opacity, so that 0.75–1 lb/gal (90–120 kg/m³) of paint normally provides adequate hiding power. Since these pigments also offer fine durability and lightfastness, they are used in such finishes as storefront enamels, pump enamels, automotive enamels, bulletin paints, *etc*. Tolidines tend to be somewhat soluble in aromatics, which must therefore be kept to a minimum. They are also not the best pigments for baking finishes, since they sometimes bronze; or for tinting colors, since they are somewhat fugitive in very low concentrations; they also bleed.

Para red is an azo pigment deeper in color than tolidine and not quite so bright. Para red offers very good coverage, so that about 1 lb/gal (120 kg/m³) gives adequate coverage. Para red is not so lightfast as tolidine and tends to bleed in oil to a greater degree. Moreover, it has poor heat resistance and cannot be used in baked coatings. Its low cost makes it attractive for bright interior finishes and some exterior finishes.

Rubine reds are bright reds, sometimes known as β-oxynaphthoic acid (BON) reds, available in both resinated and nonresinated forms. They offer good bleed resistance but only fair alkali resistance.

Lithol red is a complex organic red with very good coverage, so that 1 lb/gal (120 kg/m³) gives adequate coverage in most instances. Lithol red is bright with a bluish cast. It is relatively nonbleeding in oil but tends to bleed in water, and its durability and lightfastness are only fair. Since it is relatively low in cost, it is used in such applications as toy and novelty enamels.

Naphthol reds are arylide pigments with excellent alkali resistance and a relatively low cost. They bleed in organic solvents and are more useful in emulsion than in oil-based paints.

Quinacidone reds come in a variety of shades, ranging from light red to deep maroon and even violet. These reds offer good durability and lightfastness and high resistance in alkalies. They also tend to be nonbleeding and show good heat resistance.

Reds sometimes used include alizarine (madder lake) red for deep, transparent finishes, pyrazolone reds for high heat and alkali resistance, and a larger series of vat colors.

3.6.11.4. Violet Pigments

The demand for violets is small because they are expensive and often have poor opacity. However, some of the more frequently used violets are discussed in the following paragraphs.

Quinacridone violets are durable and offer good resistance to alkalies and to heat.

Carbazole violets offer very good heat resistance and lightfastness. They are also nonbleeding, and their high tinting strength makes them useful for violet shades.

Other violets in use include tungstate and molybdate violets for brilliant colors and violanthrone violet for high resistance and good lightfastness.

3.6.11.5. Blue Pigments

Blues not only are important as straight and tinting colors, but they are also used with other colors to produce different shades and colors.

Iron blue is a popular blue from a complex iron compound; it is also known as prussian blue, milori blue, and chinese blue. Iron blue is one of the most widely used blue pigments in the coatings industry because it combines low cost, good opacity, high tinting strength, good durability, and good heat resistance. Unfortunately, it offers very poor resistance to alkalies and cannot be used in water-based paints.

Ultramarine blue, sometimes known as cobalt blue, is popular as a tinting color. It produces an attractive reddish cast when added to whites. Ultramarine blue has poor opacity, high heat resistance, and good alkali resistance. While it can be added to latex paints, special grades low in water-soluble salts must be used. Ultramarine blue is often added to whites for extra opacity and a bluish cast, which make them look whiter.

Phthalocyanine blue is increasingly popular because of its excellent properties: It gives a bright blue color and has excellent opacity, durability, and lightfastness. In addition, it is relatively nonbleeding and produces a greenish blue shade when used as a tint. Its high chemical and alkali resistance makes it satisfactory for water-based coatings as well as all types of interior and exterior finishes. The price, though high, is not so high as to prohibit using this blue in most finishes.

Other blues sometimes used include indanthrone blue, which has a reddish cast and high resistance; and molybdate blue, which produces a very brilliant blue.

3.6.11.6. Yellow Pigments

Although *yellow iron oxide* pigments give a series of rather dull colors, they have excellent properties: relatively easy to disperse, nonbleeding, good opacity, fine heat resistance, and low cost. Since their chemical and alkali resistance is excellent, they can be used in both water- and solvent-based paints.

Excellent durability makes these oxides useful for all types of exterior coatings, and when added to white, they produce such popular shades as ivory, cream, and buff.

Chrome yellow comes in a variety of shades from very light greenish yellow to dark reddish yellow. Chrome yellow paints have good opacity, and they are easy to disperse, but they tend to darken under sunlight. Because they are lead pigments, they are gradually being phased out of use.

Cadmium yellow is largely a combination of cadmium and zinc sulfides plus barites that some in a variety of shades. These pigments offer good hiding and lightfastness when used as straight colors. They are also bright, do not bleed, bake well, and have good resistance except to acids. Since they are toxic, however, they are being phased out of use.

Hansa yellow is becoming increasingly important as bright yellows due to the elimination of chrome yellow and cadmium yellow. Hansa yellow comes in several shades, ranging from light to reddish yellow. These pigments have excellent lightfastness when used straight, but they are somewhat deficient in tints. Although their hiding power is only fair, they have excellent tinting strength, which makes them good tinting pigments, especially in water-based coatings, given their excellent alkali resistance; however, they bleed in solvents and do not bake well.

Benzidine yellow, along with hansa yellow pigments, is increasingly being used as lead-containing yellows becomes illegal. The benzidines are stronger than hansa yellows and offer good alkali and heat resistance; in addition, they are more resistant to bleeding. Since their lightfastness is inferior to hansa yellow, however, benzidine yellow pigments are unsatisfactory for exterior coatings.

Other yellows in use include nickel yellows, which offer good resistance and produce a greenish yellow; monarch gold and yellow lakes, which produce transparent metallic gold colors; and vat yellow, which offers extremely good lightfastness and resistance to heat and bleeding.

3.6.11.7. Orange Pigments

A very popular bright orange, with its reasonable cost, hiding power, brightness, and colorfastness, *molybdate orange* is nevertheless being phased out because of its lead content.

Chrome orange, a lead pigment, is also being phased out. In money value, it is inferior to molybdate orange.

Benzidine orange is bright and offers good alkali resistance and high hiding power. The pigments also offer good heat and bleeding resistance, and they can be used in both water- and solvent-based paints. Since their lightfastness is only fair, they are not the best pigments for exterior paints.

Dinitroaniline orange, a bright orange, has very good lightfastness and good alkali resistance, making it a good exterior pigment for aqueous systems. It tends to bleed in paint solvents.

Other oranges sometimes used are orthonitroaniline orange, which is lower in cost but inferior in most properties to dinitroaniline orange; transparent orange lakes, which are used for brilliant transparents and metallics; and vat orange, which is high in overall properties but also high in price.

3.6.11.8. Green Pigments

Until recently, *chrome green* was the most popular of all greens for brightness, durability, hiding power, and low cost; however, it is gradually being replaced by other greens because of its lead content. Chrome green ranges in color from light yellow green to dark blue green. It has poor alkali resistance and cannot be used in latex paints.

Phthalocyanine green is fast becoming the most important green pigment in the coatings industry. It is a complex copper compound of bluish green cast that offers excellent opacity, chemical resistance, and lightfastness. It is also nonbleeding and can be used in both solvent- and water-based coatings as either a straight color or a tint, but it is rather expensive.

Chromium oxide green is a rather dull green pigment with excellent durability and resistance characteristics; it can be used for both water- and oil-based interior and exterior paints. Chromium oxide green has moderate hiding power, and it is easy to emulsify. Its high infrared reflection makes it an important green in camouflage paints.

Pigment green B is used primarily in water-based paints because of its excellent alkali resistance, but it can also be used in solvent-based paints. It offers only fair lightfastness, so it is not satisfactory in exterior paint use. This pigment does not produce a clean shade of green, but it is satisfactory in most instances.

3.6.11.9. Brown Pigments

Most of the browns used in the coatings industry are *brown iron oxide* colors. These are essentially combinations of red and black iron oxides, offering very good coverage, excellent durability, good resistance to light and alkalies. They are suitable for both water- and solvent-based paints and either interior or exterior finishes.

Van Dyke brown is an organic brown that produces a purplish brown color. It is lightfast and nonbleeding and used primarily in glazes and stains.

3.6.11.10. Metallic Pigments

By far the most important of the metallic pigments, *aluminum* is platelike in structure and silver colored, and available in a variety of meshes and in leaving and nonleafing grades. The coarser grades are more durable and brighter, while the finer grades are more chromelike in appearance. The nonleafing grade is used when a metallic luster is required by itself or with other pigments. It is used for so-called hammertone finishes. The leafing grade is highly reflective, making it ideal for storage tanks, since it tends to keep the contents cooler. It is also very popular for structural steel, automobiles, radiators, and other products with metallic surfaces. It is used when a silver color is desired. Aluminum powder offers high opacity, high heat resistance, and excellent durability.

Bronze powders consist primarily of copper, zinc, antimony, and tin mixtures. They come in a variety of colors ranging from bright yellow gold to dark brown antique gold. Bronze powders are generally used for decorative purposes. Their opacity is poorer and their price higher than aluminum powder.

Zinc dust is assuming increasing importance as a protective pigment for metal, especially since lead is gradually being eliminated. Zinc dust is used in primers to prevent corrosion on steel when employed as the sole pigment in so-called metal filled or zinc-rich paints; and it is combined with zinc oxide in zinc dust–zinc oxide primers. Zinc dust–zinc oxide paints are satisfactory for both regular and galvanized iron surfaces. Zinc-rich paints are used with both inorganic vehicles, such as sodium silicate, and organic vehicles, such as epoxies and chlorinated rubber. Both types offer excellent rust inhibition and good weather resistance.

Lead flake is used in exterior primers, where it exhibits excellent durability and rust inhibition.

3.6.11.11. Special-Purpose Pigments

Some pigments are used for the special properties they give a coating rather than for their color or opacity. Two that have been mentioned in the metallic pigment category are zinc dust and lead flake, which are used primarily for rust inhibition. Others are discussed in the following paragraphs.

Red lead is used almost exclusively for corrosion-inhibiting metal primers, especially on large structures, such as bridges, steel tanks, and structural steel. Since this bright orange pigment has poor opacity, it is sometimes combined with red iron oxide for improved opacity and low cost. Because of restrictions on the use of lead, red lead is being phased out.

Basic lead silicochromate is also a bright orange pigment used primarily as a rust inhibitor for steel structures. Because of its low opacity, this pigment is combined with other pigments to produce rust-inhibiting topcoats in different colors.

Lead silicate is used mainly in water-based primers for wood, where it reacts with tannates to prevent them from bleeding and discoloring succeeding paint coats. Lead silicate may be eliminated from home use because of restrictions on the use of lead.

Zinc yellow is a hydrated double salt of zinc and potassium chromate used principally in corrosion-inhibiting metal primers. It is becoming one of the few pigments permitted for use on steel connected in houses or apartments. The pigment is greenish yellow and offers poor opacity.

Basic zinc chromate possesses properties somewhat similar to those of zinc yellow. This pigment is used in metal pretreatments, especially in the well-known "wash primer" government specification for conditioning metals, where it promotes adhesion and corrosion resistance in steel and aluminum.

Cuprous oxide is a red pigment used almost exclusively in autifouling ship bottom paints to kill barnacles.

Antimony oxide is a white pigment used almost exclusively in fire-retardant paints, where it has been very effective, especially in combination with whiting and chlorinated paraffin.

4

Lead Pigments

4.1. HISTORY OF LEAD PIGMENTS

The Book of Job mentions lead as a writing material. In one of his replies to Bildad, Job exclaims:

> Oh that my words were now written! Oh that they were printed in a book! That they were graven with an iron pen and lead in the rock for ever. For I know that my redeemer liveth . . . Job 19:23–25

Commentators disagree about how this writing was done; some maintain that the characters were engraved on a lead plate with an iron stylus, whereas others believe that the stylus was used to engrave the rock, then molten lead was poured into the etched marks.

After the Pharoah's chariots had been engulfed by the Red Sea, Moses and the children of Israel sang an anthem of thanksgiving, "Thou didst blow with they wind, the sea covered them: They sank as lead in the mightly waters" (Exod. 15:10).

In the time of Zechariah (a century later), lead weights were in use: "And behold, there was lifted up a talent of lead. . ." (Zech. 5:7). Ecclesiassticus said of King Solomon, "Thou didst gather gold as tin, and didst multiply silver as lead" (Eccles. 47:18).

Lead ores are widely distributed in nature and easily smelted. The Babylonians engraved inscriptions on thin plates of metallic lead, and the Romans used lead extensively for water pipes, writing tables, and coins. Unfortunately, they also used lead for cooking utensils, and lead poisoning was an all-too-frequent result. A few small lead nuggets, some of which are believed to be of pre-Columbian origin, have been found in Peru, the Yucatan, and Guatemala.

Theophrastus (Weeks, 1956) described the manufacture of "ceruse" (basic lead carbonate or white lead) in *History of Stones* (372–287 B.C.):

> Lead is placed in earthen Vessels, over sharp Vinegar, and after it has acquired some Thickness of a kind of Rust, which it commonly does in about ten Days, they open the Vessels, and scrape it, till it is wholly dissolved; what has been scraped off they then beat to Powder, and boil for a long Time; and what at least subsides to the Bottom of the Vessel is the Ceruse.

By the time of Dioscorides (first century A.D.) the process had undergone little or no change. Dioscorides also described white lead, distinguishing it from cinnabar, and mentioned its use for painting and decorating walls.

Marcus Vitruvius, architect and engineer under the Emperor Augustus, was familiar with the toxicity of lead and observed that laborers in the smelter had pale complexions because of their prolonged exposure to lead dust and vapor. Pliny (A.D. 23–79) warned about the hazards of lead. However, the Romans continued to add lead acetate to wine to enhance its taste because they failed to recognize the toxicity of lead. Hippocrates and Galen noticed that slaves who worked in the mines were suffering and dying from acute neurologic and kidney impairments.

J. P. de Tournefort, who visited the Levant in 1700, wrote:

> Spiphanto, in days of yore, was famed for its rich Gold and Silver Mines . . . Besides the Mines aforesaid, they have plenty of Lead; the Rains make a plain discovery of this, go almost where you will throughout the whole Island. The Ore is greyish, sleek, and yields a Lead like Pewter.

The lead mines (Weeks, 1956) of Missouri (formerly known as the lead mines of Louisiana) were discovered in 1720 by Philip Francis Renault and M. La Motte, who afterward mined them by the open-cut method. The famous Burton mine, discovered more than half a century later, was inefficiently mined by the Spaniards. In 1797, Moses Austin of Connecticut sank the first shaft, installed a reverberatory furnace, and manufactured shot and sheet lead. When the United States purchased the vast region formerly known as Louisiana from France in 1803, the lead industry was already well developed.

About 1819, Henry R. Schoolcraft (Weeks, 1956) visited all the lead mines in the Missouri region, exploring the minerals and geological structures. He found the lead mainly in the form of the sulfide galena. Zinc sulfide, or sphalerite, was abundant, but mining techniques had not been developed. This tristate area (Missouri, Kansas, and Oklahoma) has since become one of the world's leading sources of both lead and zinc.

Common lead ores (Hurlbut, 1966) include galena, cerussite, angelsite, phosgenite, pyromorphite, mimetite, vanadinite, crocoite, and wulfenite;

Table 4.1 lists lead pigments (Weast, 1978; United States of America before the Federal Trade Commission, 1953).

4.2. DEVELOPMENT OF WHITE LEAD PIGMENTS

White lead was first manufactured in the United States at the beginning of the nineteenth century, and its principal use was as a paint pigment (Minerals Resources, 1928). Smith (1991) reported the following figures in the records of Civil Action No. 87-2799 T, U.S. District Court, District of Massachusetts; this information is documented in *Minerals Yearbook*: for 1929 (1928), 90%; for 1932–1933 (1929), 90%; for 1930, 89.6%; for 1934 (1933), 93%; and for 1935 (1934), 95%. By the beginning of the twentieth century, white lead was already widely promoted for use in both interior and exterior house paint. White lead production increased in earnest in the late part of the nineteenth century, and white lead production peaked in the early-to-middle 1920s. It steadily declined thereafter except for a brief period in the late 1930s and early 1940s when the lead industry redoubled its efforts to promote white lead for paint (*Minerals Yearbook*, 1955; *Dutch Boy Painter Magazine*, 1939, 1940, 1943). By the beginning of the twentieth century, other white pigments were being developed as alternatives to white lead. Calls for the elimination of lead from paint, largely due to its toxicity, began in the nineteenth century and increased in intensity during the first few decades of the twentieth century (Tanguerel des Planches, 1948; Goodell, 1882; Oliver, 1914). No later than 1920, nonlead pigments were available for interior surfaces as an effective alternative to white lead; effective alternatives to white lead for exterior surfaces could have been available by no later than 1930 (*Minerals Yearbook*, 1928). Despite the availability of alternative pigments, white lead was promoted during the 1930s and 1940s. Although in the 1950s, interior house paints declined in their lead content, high-volume leaded house paint was sold without proper labeling and used on interior surfaces (*Chemical Week*, 1954).

The first substitute for white lead as a paint pigment was zinc oxide, first produced in the United States in the mid-1800s. Its use as a pigment increased in the nineteenth century, so that by 1988, zinc oxide production in the United States was approximately one-fourth that of white lead. Studies in the late nineteenth and early twentieth centuries comparing zinc oxide to white lead for interior surfaces confirmed that zinc represented an effective alternative (Goodell, 1882; Oliver, 1914; Mineral Industry, 1903; Gardner, 1917; International Labor Office, 1927). Zinc oxide and lithopone were often described as the competitive alternative to white lead as a paint pigment. Lithopone, a

Table 4.1. Lead Pigments

Pigment	Color	Formula/Comments
Lead acetate	White	$Pb(C_2H_3O_2)_2$
Dry white lead, basic carbonate	White	$2PbCO_3 \cdot Pb(OH)_2$
Dry white lead, basic sulfate	White	$PbSO_4 \cdot PbO$
White lead in oil	White	Basic carbonate in linseed oil
Litharge (lead oxides)	Yellow black black red	PbO—yellow Pb_2O (sub-oxide)
Red lead	Orange	Pb_2O_4—red Pb_3O_4
Blue lead	Blue	Basic lead sulfate made by a fuming process
Orange mineral	Orange	Further processing of red/white lead
Grinder's lead in oil	White	White lead in oil
Pig lead	Gray	Metallic lead
Lithopone	White	Precipitated zinc sulfide (30%) barium sulfate (70%)
Lead chromate/oxide	Orange	$PbCrO_4 \cdot PbO$
Basic lead chromate	Red	$Pb_2(OH)_2 2CrO_4$
Lead chromate	Yellow	$PbCrO_4$
Lead chromate (Green)	Green	Coprecipitated lead chromate/alkaline ferroferri-cyanide
Basic lead silico-chromate	Red orange	$PbCrO_4$ on silica particle
Basic silicate white lead	White	$2PbSiO_3 \cdot Pb(OH)_2$ on silica particle

mixture of zinc sulfide and barium sulfate was invented in France around 1875; the mixture was well suited for interior surfaces, and when mixed with zinc oxide and calcium carbonate, it could be used on exterior surfaces as well. In addition to its nontoxic qualities, lithopone was cheaper to use in paint than white lead. By 1920, New Jersey Zinc had introduced a light, stable lithopone that provided good quality at low cost (Protective and Decorative Coatings, 1941). In 1924, Glidden introduced Zinc-O-Lith®, a lithopone product, as a new kind of house paint. In European countries where the use of white lead in paint was prohibited, a product similar to Zinc-O-Lith®, which produced the same results at a lower cost, was used (Haynes, 1945–1956). Lithopone production surpassed white lead in 1926 and peaked in 1929 (Lead and Zinc Pigments, 1955). During this period, both types of pigment were used primarily in paint (85% of all white lead and 75% of all lithopone), but lithopone cost one-third to one-half less than white lead.

The third principal substitute for white lead was titanium dioxide (*Minerals Yearbook*, 1920–1935). The production process was first developed in Norway; the first U.S. commercial production of titanium white began in 1918 by Titanium Products (Haynes, 1945–1956). In 1920, National Lead purchased an interest in Titanium Products and acquired the remaining interest in the company by the 1930s. In 1931, DuPont began producing titanium dioxide, and by the mid-1930s, National Lead and DuPont were the major producers of the pigment. In 1924, the Bureau of Standards issued proposed master specification for a lead-free exterior white titanium and zinc paint (*Minerals Yearbook*, 1924). Tests comparing titanium dioxide to white lead demonstrated the former to be a technologically competitive pigment (Tyler *et al.*, 1929; Campbell Paint Brochure, 1924). Moreover, by 1929 titanium pigments had assumed an important role in European countries that had legislatively restricted the use of white lead in paints (*Minerals Yearbook*, 1929).

4.3. WHITE LEAD PIGMENT PRODUCTION IN THE UNITED STATES

During the first seven decades of the twentieth century, the U.S. white lead industry was made up of a relatively small number of white lead manufacturers. By 1910, lithopone was recognized by a number of producers and was at all times dominated by National Lead. With specific adjustments for periods of production, the principal white lead producers consisted of National Lead, Eagle-Picher Lead, Sherwin–Williams, Glidden, and International Smelting and Refining. Other white lead producers, such as E. I. du Pont de Nemours (*Minerals Yearbook*, 1921–1928), W. P. Fuller, and John R. MacGregor Lead (*Minerals Yearbook*, 1945–present) maintained relatively small positions within the industry (United States of America before the Federal Trade Commission, 1953). White lead imports never accounted for a significant percentage of overall domestic consumption (*Minerals Yearbook*, 1920). Market shares for white lead shipments reported by the Federal Trade Commission (United States of America before the Federal Trade Commission, 1953) generally reflect market shares for earlier periods.

The first U.S. white lead factory was established in 1804 by Samuel Wetherill and Sons in Philadelphia. By the mid-1800s, there was rapid growth in white lead manufacturing in the United States, which included the Eagle White Lead around 1843 (*Engineering and Mining Journal*, 1943). In 1891, National Lead was formed by the acquisition of the physical properties and stock of a substantial number of then existing white lead companies (National Lead v. Federal Trade Commission, 1953). Among the acquisitions by National Lead during this period was the 1906 acquisition of United Lead, which gave National Lead approximately 85% of U.S. white lead production (Ingalls, 1908). In 1916, Eagle White Lead merged with the Picher Lead to form the Eagle–Picher Lead (*Engineering and Mining Journal*, 1943). Eagle–Picher and National Lead were positioned as the two major supply sources for white lead in the United States (*House that Joyce Built*, 1949). Thereafter, the only substantial entry of white lead producers into the industry consisted of paint companies (*e.g.*, Sherwin–Williams and Glidden) that wished to avoid depending on the two big producers and companies introducing new production technology (*e.g.*, Anaconda) (*National Lead v. Federal Trade Commission*, 1953). Thus, by 1914, Sherwin–Williams started a lead-corroding operation in Chicago, Illinois, and in 1920 acquired the Acme White Lead and Color Works in Detroit, Michigan (*Fortune*, 1935). In 1924, Glidden acquired Euston Lead. Then in the early 1920s, Anaconda Lead Products began producing white lead using an electrolytic process at its plant in East Chicago, Indiana (*Minerals Yearbook*, 1922).

In 1921, the following companies were producing white lead (*Minerals Yearbook*, 1921):

- National Lead and subsidiaries
- Eagle–Picher Lead and subsidiaries
- Euston Lead
- Sherwin–Williams
- Anaconda Lead Products
- W. P. Fuller
- Wetherill and Bros.
- E. I. du Pont de Nemours.

National Lead had the greatest number of producing plants, which is consistent with data showing that National Lead was and remained the largest producer of white lead in the United States.

In 1933, the following companies were producing white lead (*National Lead v. Federal Trade Commission*, 1953):

- National Lead

- Eagle–Picher Lead
- Sherwin–Williams
- Euston Lead (acquired by Glidden in 1924)
- Anaconda Lead (International Smelting and Refining)
- Wetherill and Bros. (acquired by National Lead in the 1930s).

In 1945, the following companies were producing white lead (*Minerals Yearbook*, 1945):

- National Lead and subsidiaries (multiple plants)
- Euston Lead (one plant)
- John R. MacGregor Lead (one plant)
- W. P. Fuller (one plant)
- International Smelting and Refining (one plant)
- Eagle–Picher Lead (one plant)
- Sherwin–Williams (one plant).

In 1946, the Federal Trade Commission charged the following companies with effecting monopolies, lessening competition, and restraining trade in the sale of lead pigments, including white lead (*National Lead v. Federal Trade Commission*, 1953):

- National Lead
- Eagle–Picher Lead
- Sherwin–Williams
- International Smelting and Refining
- Glidden.

During the 1930s and early 1940s, these companies accounted for practically the entire lead pigment production in the United States (*National Lead v. Federal Trade Commission*, 1953). There is no indication in government publications, journal articles, or other texts that there were any other companies producing a substantial amount of white lead during this period or afterward through the 1950s (National Paint, Varnish, and Lacquer Association, 1951). In 1946, International Smelting and Refining sold its only white lead producing plant to Eagle–Picher Lead. In 1949, Glidden produced about 5% of the white lead carbonate and white lead in oil that had been produced in the United States. Sherwin–Williams and Eagle–Picher ceased producing white lead sometime prior to 1953. Glidden sold its white lead plant in the late 1950s. National Lead continued producing white lead at least into the 1960s.

4.4. SUMMARY

Table 4.2.   Chronology of Lead Pigment Development

Year	Major Event
20,000 B.C.	Human beings first exposed to lead during the Bronze Age.
1000–4000 B.C.	Old Testament mentions the use of lead.
2000–4000 B.C.	Romans and Babylonians used lead.
372–287 B.C.	Theopharstus described the manufacture of ceruse (basic lead carbonate).
100 B.C.	Pliny warned about the hazards of lead.
	Hippocrates and Galen noticed that slaves working in lead mines suffered and died from acute neurologic and kidney impairment.
	Marcus Vitruvius observed that laborers in the smelters had pale complexions.
1700	J. P. de Tournefort wrote about lead mines in the Levant area.
1720	Phillip Francis Renault discovered lead mines in Missouri.
1797	Moses Austin sank the first shaft, installed a reverberatory furnace, and manufactured lead shot and sheet.
1804	White lead first manufactured in the United States.
1819	Henry R. Schoolcraft visited all lead mines in the Missouri region (Missouri, Kansas, and Oklahoma), which became one of the world's largest lead mining areas.
1900	White lead paint was used on exterior/interior surfaces in houses.
1920s	U.S. white lead production peaked; substitutes were available.
1930s	U.S. white lead production declined.
1940s	U.S. white lead producton increased again.
1950s	Lead pigments declined in house paint, which was sold without proper labeling.
1960s	Most lead companies sold their interests in lead manufacturing.
1973	Lead content in paint reduced by U.S. Congress to 0.5% by weight for residential uses.
1978	Lead content in paint reduced by U.S. Congress to 0.06% by weight for residential uses.

5

Health Problems Associated with Lead-Based Paint

5.1. GENERAL TOXICOLOGY OF LEAD-BASED PAINT

5.1.1. History of Lead Poisoning

The archeological record of human remains and archeological geochemistry show that human beings were first exposed to lead 20,000 years ago in the Bronze Age as they began to exploit various metals. Human bones, ice, and sediment cores from that period show some lead contamination associated with using lead. This contamination increased dramatically starting about the fourth century B.C. with the use of lead and silver in Greece and then in Rome. Since silver and lead are frequently found together in the same ore, silver extraction released a great deal of lead into the environment. The adverse effects of lead on the human body were reported by the fathers of medicine, Hippocrates and Galen, when they noticed that slaves working in mines were suffering and dying from acute severe neurologic and kidney impairments. During the Roman Empire (A.D. 100 onward), lead was used in plumbing, ceramic glazes, *etc*. In the first century B.C., Pliny warned about the hazards of lead. However, the Romans continued to add lead acetate to wine to enhance its taste because they failed to recognize the toxicity of lead.

5.1.1.1. Lead Poisoning from 1000–1880

In the seventeenth century, lead was outlawed as an additive to wine in France. In the eighteenth century, Benjamin Franklin wrote about the hazards of lead, particularly for printers casting their own lead type. The effects of lead at lower doses were recognized in the mid-1700s with the growth of the English pottery works in Staffordshire. There, workers routinely ingested and inhaled lead dust during the glazing process. By the middle and late nineteenth century, there began to be concern about the adverse effects of the relatively low-level exposure of women and children to lead dust brought into the home on the clothing of pottery workers. For perhaps the first time, the medical and industrial communities recognized that lead outside the workplace could be a problem and that children were very sensitive to lead.

5.1.1.2. Lead Poisoning 1897–1922

At around the turn of the century, there was increasing concern for the nonoccupational hazards of lead. In Australia, pioneering works by A. J. Turner in 1897 and J. Lockhart Gibson in 1904 concluded that lead paint in porch railings was responsible for lead poisoning in children, and in 1922, the state of Queensland, Australia, banned the use of lead paint for certain dwelling surfaces (Landrigan, 1991).

5.1.1.3. Lead Poisoning from 1920–1943

By 1917, with Blackfan's article on "Lead Poisoning in Children" in the *American Journal of Medical Science*, it was well established in the United States that lead paint poisoning was a frequent cause of death among children. In the early 1920s, medical literature contained reports of childhood lead poisoning that resulted from ingesting paint from window sills, porches, furniture, and toys. During the first quarter of the twentieth century, legislation aimed at reducing both occupational and nonoccupational exposure to lead and lead paint was enacted throughout the world. Scientific and medical evidence that lead paint was the major source of childhood lead poisoning continued to grow dramatically throughout the 1930s. The Baltimore health department routinely began to investigate cases of childhood lead poisoning and reported 135 cases, including 49 deaths, between 1930 and 1941. In 1935, it established a free blood level laboratory service for physicians and hospitals.

Byers and Lord investigated the long-term health effect in children who had experienced lead poisoning in the 1940s. In 1943, Byers reported on 20 children who had recovered from acute lead intoxication, noting that 19 of the 20 had behavior disorders or learning disabilities. This marked the beginning of modern lead intoxication studies, which continues at present.

5.1.2. Effects of Lead Poisoning

Virtually no part of the body is immune from the effects of lead. Lead in the body disrupts energy metabolism at the cellular level, interferes with neural cell function in the brain, disrupts the formation of heme and in the nervous system, inhibits communication and decreases nerve conduction velocity. Even at relatively low exposures, lead's neurotoxic effects can cause decreased intelligence, short-term memory loss, reading under-achievement, impairment of visual-motor function, loss of auditory memory, poor perceptual integration, poor classroom behavior, and impaired reaction time.

5.1.2.1. Effects on the Body

The mechanism by which lead affects the human body is extremely complex, as demonstrated by Landrigan and Needleman (1991) in Civil Action No. 87-2799-T, U.S. District Court for the District of Massachusetts in 1991. On an atomic level, lead has no biologic value and competes with metals that are essential to the human body, such as zinc, iron, and calcium. When lead inserts itself into the binding site of proteins, such as enzymes, in place of these essential metals, it distorts the biomolecules and their functions. For example, in certain membrane channels, lead occupies calcium's place, thereby blocking the flow of calcium within cells or across cell membranes. Since calcium is a very important regulator, this lead–calcium interaction affects memory storage and cell differentiation in the nervous system.

Another molecular mechanism of lead toxicity involves lead blocking the essential transfer of oxygen from cytochromes in cells, which is necessary for cells to continue living and functioning. A third mechanism of lead toxicity involves lead's ability to inhibit the biosynthesis of heme, thus depleting cells of heme for cytochromes and accumulating biologically active heme precursors. A fourth mechanism of lead toxicity affects the regulation of DNA transcription.

5.1.2.2. Effects on the Brain

Lead inhibits certain neuronal pathways in the brain and disinhibits others. As a result, the system is less able to store information, draw on past information or memory, or inhibit responses to environmental stimuli. Lead effects the peripheral nervous system, the nerves that control the muscles and organs of the body outside the brain, by stripping off the coating, known as myelin, that ensheathes the system. This disrupts the propagation of electrical impulses to peripheral nerves and results in decreases in muscle strength and at high doses, the clinical syndromes known as wrist drop and foot drop. Lead affects the cells in the kidneys that regulate the transfer of electrolytes by accumulating in the nucleus of those cells, which can result in changes in the secretion of hormones from the kidney and in the activation of vitamin D, with far-reaching endocrinologic effects. Lead affects the liver by inhibiting heme synthesis, which results in the depletion of heme-containing enzymes, including the hepatic-mono-oxygenase enzymes. Lead also has an impact on the immune system as a result of its affect on the spleen, which is a conditioning organ for circulating cells in the immune system. Lead affects the hematopoietic system by interfering with the synthesis of hemoglobin and heme and eventually with the availability of heme for hemoglobin. This results in anemia and increased red cell fragility. Lead affects bone by substituting for calcium in the activation and function of the calciotropic hormones in the hormonal regulatory system of bone; lead also inserts itself into the bone matrix, just as calcium does. Lead affects both male and female gonads through steroid hormone receptors, which results in decreased activity of enzymes that process testosterone and other androgens. There is also evidence that lead is associated with increased blood pressure and hypertensive heart disease through its neurogenic, renal, and endocrinologic effects. Finally, lead stored in women's bones can be mobilized during pregnancy and passed on to unborn children; lead can also be re-mobilized during the postmenopausal period when bones are decalcifying.

5.1.3. Occupational and Nonoccupational Hazards of Lead Poisoning

5.1.3.1. Occupational Hazards

Medical and scientific observations of the toxicity of lead date back to antiquity (Lin-Fu, 1985). Lead poisoning was probably one of the earliest occupational diseases and one of the earliest to be recognized and diagnosed. With the increased use of lead in more recent times, the problem of occupational lead poisoning has increased. In the eighteenth century, Benjamin Franklin wrote about the hazards of lead among printers and others, and in 1839, Tanguerel des Planches produced a treatise on the problem and its causes (Tanguerel des Planches, 1839). This treatise not only provided an accurate and elaborate account of a large number of patients with industrial lead poisoning, including white lead workers and painters, but it also called for the substitution of zinc oxide in place of white lead in paint. Tanguerel's treatise was well known in the medical community and in this country in the early twentieth century and often cited as authoritative (Blackfan, 1917; Holt et al., 1923).

5.1.3.2. Nonoccupational Hazards

The subject of nonoccupational lead poisoning, including childhood lead poisoning, began to receive more serious attention during the latter part of the nineteenth century. In 1887, there were a series of lead-related deaths in Philadelphia, especially among children, from ingesting chrome yellow used on bakery buns by certain bakers (Glenn, 1889; Stewart, 1895). In 1897, Turner reported an outbreak of childhood lead poisoning in Queensland, Australia (Turner, 1897). The cause of these lead poisonings eluded Turner and his colleagues for some time, but by 1904, leaded house paint was identified as the source (Gibson, 1904; Turner, 1908). The early discoveries by Gibson, Turner, and their colleagues were reported in the United States as early as 1914, and in a published report, they were compared to a similar case of childhood lead poisoning in Baltimore (Thomas *et al.*, 1914). Three years later, Kenneth Blackfan, one of the great luminaries in U.S. pediatrics, published a review article on the subject of childhood lead poisoning and its association with lead paint (Blackfan, 1917).

The discoveries by Turner and Gibson between 1897 and 1904, and their dissemination in the United States by Blackfan in 1917, are the cornerstones of medical knowledge about childhood lead poisoning and its association with lead paint. The important medical and scientific research that followed has reinforced these basic observations and added to this body of medical knowledge by improving diagnostic techniques and discovering much more about the neurologic outcomes of lower-level lead exposure. However by 1917, the medical community was generally aware of the disease of childhood lead poisoning and its association with lead paint, and by 1917, it was well established in the United States that lead paint poisoning was a frequent cause of death in city pediatric wards. By the early 1920s, childhood lead poisoning was recognized as a common, and often fatal, childhood disease. Researchers identified window sills, porch railings, children's furniture and toys as major sources of lead paint. In 1922, the state of Queensland, Australia banned the use of lead paint for certain dwelling surfaces. Other countries that either banned the use of lead paint indoors or severely restricted children's contact with it were Tunisia (1922), Greece (1922), Great Britain (1926), Sweden (1926), Belgium (1926), Poland (1927), Spain (1931), Yugoslavia (1931), and Cuba (1934); similar bans had been in effect in France since 1917 (Williams *et al.*, 1952; Cushing, 1934; Ross *et al.*, 1935). Although physicians and scientists did not begin to investigate the long-term health effects of lead poisoning in children who survived until the 1940s, there is no question that lead-based paint in the home was widely recognized as a serious health risk to children as early as 1917.

5.2. EFFECTS OF LEAD POISONING ON CHILDREN

The accepted index of lead in the human body is the chemical measure of lead in the blood. The federal definition of lead toxicity is a blood lead concentration greater than 25 µg/dl, but this will shortly be lowered to 10 µg/dl. When the blood lead exceeds this level, it causes damage to the body, particularly the brain. Data indicate that 17% of all children, regardless of race or socioeconomic status, have blood lead levels in the toxic range. Over 50% of black children in poverty enter the first grade with elevated blood lead levels considered neurotoxic. The harmful effects of lead, moreover, are directly proportional to the level of lead in the blood; thus, over 40 µg/dl is considered unequivocally toxic; over 50 µg/dl is considered urgent; and over 70 µg/dl is considered a medical emergency.

5.2.1. General Effects

Lead poisoning causes serious neuropsychologic damage in children, characterized by reduced intelligence, learning disabilities, aberrant behavior, brain dysfunction, hyperkinesis, and attention deficit disorder; there is also evidence of regression and loss of acquired behaviors and skills.

The number of related articles and medical knowledge about the problem of childhood lead poisoning continued to grow throughout the 1920s. The issue appeared in standard pediatric texts (Holt, 1923), and by the mid-1920s numerous articles had been published describing lead exposure, including lead paint, as a serious hazard to infants and children (Ruddock, 1924). After 1926, reports also began to appear in government publications and insurance company reports (Hoffman, 1927). Related literature continued to grow throughout the 1930s, with reports of numerous cases of children afflicted with lead poisoning from house paint (Ruddock, 1932). In 1933, Jarvis Nye published an important monograph that linked lead paint exposure to chronic nephritis among children in Australia (Nye, 1933). In the same year, McKhann and Vogt reported on 89 cases of children with lead poisoning at Children's Hospital, Boston, between 1924 and 1933; they recognized the continuing presence of lead paint as a source of exposure (McKhann et al., 1933).

This research culminated in 1943 with the investigation and report by Byers and Lord on long-term effects of childhood lead poisoning (Byers et al., 1943). Their work represented the beginning of the modern age of lead intoxication studies and provided the historic analogue for modern research and medical knowledge on neurologic impairment from low-level lead exposures.

Scientific evidence has established that lead is a potent toxicant that may have irreversible effects on many organ systems. At high doses, it causes the death of brain cells, while at lesser doses it causes disturbances in membrane function that interfere with the synthesis of essential proteins in nerve cells, among other things (see "Effects on the Body"). In the human body, lead is a potent poison that can affect individuals in any age group, but it is particularly harmful to children because it can impair normal brain development. Moreover, children are more exposed to lead than adults because of their normal hand-to-mouth activity. Once absorbed, lead is stored primarily in bone and to a lesser degree in the kidneys and brain, while small quantities circulate in the blood. Lead's persistence in the body is unequalled by virtually any other toxin except possibly dioxin. Lead's "half-life" in bone—the time it takes half of a given dose to be removed—exceeds 20 years. As a result, even small amounts of lead accumulate in the body and can cause effects that endure long after exposure ends.

Because lead causes neurologic damage at doses that do not cause overt toxicity, blood lead levels are used to identify children with dangerous amounts of lead (see Chapter 1). Another measurement of body lead levels is the free erythrocyte protoporphyrin (EP or FEP) test, which measures the level of EP in the blood. The EP level reflects an impaired ability of the body to produce heme, the "business end" of the hemoglobin molecule, usually due to the presence of lead. Blood lead levels reflect a child's immediate past exposure to lead as well as lead previously absorbed by bone and now being released into the bloodstream. Thus a child may continue to have toxic levels of lead in his/ her blood long after a lead source has been removed.

Treatment for children with lead poisoning involves removal of the source of lead and careful clinical and laboratory surveillance of the child. Because lead is excreted so slowly, many children with elevated levels of blood lead require chelation therapy to lower their lead levels. This involves giving the child one of several different chemicals intravenously, through injection, or orally, depending on the chemical used, which binds with the lead, causing the body to excrete it more rapidly.

5.2.2. Epidemiologic Studies

Epidemiologic studies have clearly documented the serious long-term effects of exposure to lead in childhood. These studies, from Europe, Australia, and New Zealand, as well as from the United States, have established that even small amounts of lead are associated with lower intelligence quotient scores and other significant behavioral changes in children (Needleman *et al.*, 1990). Studies in 1979 reported that first- and second-grade children without symptoms of plumbism but with elevated detin levels, had deficits in psychometric intelligence scores, speech and language process, attention, and classroom performance (Needleman *et al.*, 1979). When those same children were studied in the fifth grade, the children with high dentin lead levels had lower intelligence quotient scores, needed more special academic services, and had a significantly higher rate of failure in school than other children (Bellinger *et al.*, 1986). When some of these subjects were reexamined as young adults, those with high dentin lead levels had a markedly higher risk of dropping out of high school as compared to those with low dentin lead levels. Higher lead levels in childhood were also significantly associated with lower class standing in high school, increased absenteeism, lower vocabulary and grammatical-reasoning scores, poorer hand–eye coordination, longer reaction times, and slower finger tapping. Some medical researchers have concluded that exposure to lead in children is associated with deficits in central nervous system functioning that persist into young adulthood.

Some medical researchers believe, to a reasonable medical and scientific certainty, that the effects of lead on the brain are permanent and irreversible, and they seem to worsen with time. This is reflected not only in lower intelligence quotient scores, but in a decreased ability to function in every day life. There is growing evidence that children exposed to lead have attention deficit disorder and are hyperactive. Although only about 3% of children generally suffer attention deficit disorder, 55% of children poisoned by lead paint display signs of this disorder. Such children are at high risk for aggressive and antisocial behavior in later life.

The consequences of exposure to lead are more serious the younger the child is at the time of exposure. Evidence indicates that young children are more sensitive to injury from lead because their nervous systems are still developing and thus more subject to permanent damage.

There are other serious health effects of lead exposure: Blood lead levels are significantly related to stature, and a blood lead level of 26 μg/dl has been associated with a 3% reduction in height. Other studies have found that lead was positively associated with hearing loss. A number of recent studies have shown a positive relationship between lead exposure and blood pressure. There is also a danger that the child of a mother exposed to lead poisoning when she was a child will have excess blood lead levels, since most lead in the body is deposited in bone, where it remains relatively inert for long periods of time. But during pregnancy, hormonal changes may mobilize these lead stores, creating an endogenous source of fetal exposure.

5.3. SUMMARY

The 1988 Agency for Toxic Substances and Disease Registry (ATSDR) Report to Congress estimates that there is still 5 million tons of lead in the

United States in household paint. Seventy percent of the houses built before 1960 have lead-based paint on interior and exterior surfaces. The ATSDR further estimates that there are 6 million homes with leaded surfaces that have decayed or deteriorated and in which 2 million young children presently reside. Ingesting lead from chipped, peeling, or chalking paint places these children at extremely high risk of lead poisoning.

Table 5.1 summarizes the history of lead toxicity.

Table 5.1. Chronology of Lead Toxicology

Date	Major Events
20,000 B.C.	Human bones from the Bronze Age found to contain lead.
100 B.C.	Romans used lead in plumbing pipes and cooking utensils; Romans used lead acetate to sweeten wine; lead poisoning was suspected in Romans.
1700–1880	Lead use in wine was banned in France
	Benjamin Franklin wrote about the hazards of lead in lead type used for printing.
1887	A series of deaths were reported in Philadelphia, Pennsylvania, related to chrome yellow used on buns by certain bakers.
1897	Turner reported an outbreak of childhood lead poisoning in Queensland, Australia, attributed by 1904 to houses painted with lead-based paint.
1904	Dr. J. Lockhard Gibson established that lead paint on porch railing was responsible for lead poisoning in children in Queensland, Australia.
1917	Blackfan published "Lead Poisoning in Children" in the *American Journal of Medicine*.
1930s	Scientific and medical evidence of lead poisoning in children grew.
1935	Baltimore City Health Department established a blood-testing laboratory for physicians and hospitals.
1943	Byers reported behavior disorders in children due to lead poisoning.
1973	The U.S. Congress passed legislation requiring all residential paint to contain less than 0.50% lead by weight.
1978	The 1973 limit on lead in paint was reduced to 0.06% by weight.

6

Field Detection Methods

6.1. DETECTING LEAD IN PAINT

Environmental regulations covering abatement and removal of lead-containing paint in residential structures have led to several studies on procedures for measuring lead concentrations in paint films (Luk *et al.*, 1990). As discussed in Chapter 1, current federal regulations require abatement of painted surfaces in HUD-assisted housing having lead concentrations comparable to 1 mg/cm^2 (*Federal Register*, 1988).

Although standard laboratory methods for measuring lead concentration in paint films are available, (ASTM D 3335, 1989; NIOSH, 1984) questions remain about the precision and accuracy of field methods. Field methods are desirable because of the large number of existing painted surfaces to be tested. Some of the typical residential structure surfaces to be measured are shown in Figure 6.1. Two general types of field measurement methods are being used to detect or measure lead in paint: portable X-ray fluorescence (PXRF) and chemical spot tests (McKnight *et al.*, 1989).

6.2. PORTABLE X-RAY FLUORESCENCE

A material is said to fluoresce if it emits radiation as the result of absorbing higher energy radiation from some remote source. Each element has a characteristic X-ray fluorescence spectrum that is essentially independent of the composition of the material; that is, the characteristic lead X-ray fluorescence spectrum is the same for red lead pigments as for white lead pigments.

Commercially available PXRF instruments used to measure lead in paint are hand-held devices shaped like a small rectangular box. They typically

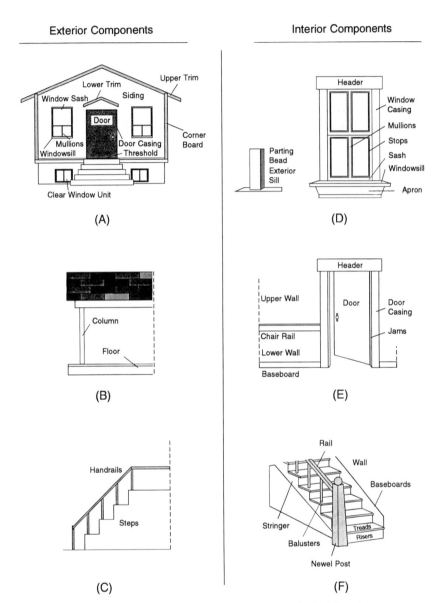

Figure 6.1. Residential structures and components for lead detection.

weigh about 12 lb (a few kilograms) and are about 1 ft long (about 0.333 m). The part of the device that comes into contact with a painted surface has a rectangular face on the order of 3.14 in. $\times$ 3.94 in. (0.08 m $\times$ 0.1 m). The detector component is hand held and placed on a sample to take measurements. The sample can be vertical, horizontal, or otherwise oriented, since the detector component is very portable. Many of the devices have a separate read-out and power unit. To take a measurement, the instrument is held against a surface, then a shutter is opened using a trigger or other device to expose the surface to the source. As long as the shutter is open, X-rays from the source bombard the specimen and excite atomic electrons. As the electrons return to their ground state, X-rays produced by the excited atoms, as well as some scattered X-rays, come into contact with the detector and are analyzed to measure lead concentration. Individual measurements typically take about 1 minute. The cost of a measurement varies from \$1–\$50, as reported to the author by testing companies. A large number of measurements cost less per measurement than an individual measurement, since the operator and equipment have to travel to the test site, and start-up time is distributed over several samples.

Most PXRF instruments used to detect lead in paint films bombard the films with X-rays emitted by radioactive cobalt (Co^{57}). Most of these X-rays have sufficient energy to excite lead k-shell electrons (k absorption edge is 88 keV) to penetrate many paint layers. The X-rays produced as a result of exciting k-shell electrons also have sufficient energy so that they are not readily attenuated by atoms of other elements commonly found in paint films. Thus, measurements where k-shell electrons are excited are rather sensitive to the depth at which lead atoms are present. However, because of the high energy of the Co^{57} X-rays, they are not readily attenuated by elements likely to be present in paint and many building materials that can be scattered by atoms in materials well below the paint films. Thus, the substrate and other underlying materials can affect the measurement.

On the other hand, some portable devices use radioactive sources that emit X-rays that do not have sufficient energy to excite k-shell lead electrons but do have sufficient energy to excite l-shell lead electrons. The characteristic lead X-rays produced in this process have much lower energies than those produced when k-shell electrons are excited and can be attenuated by elements commonly found in paint. Thus, the depth of the lead atoms below the surface affects the measurement result (McKnight *et al.*, 1990).

There are two general types of commercially available instruments for measuring lead in paint films that are based on exciting k-shell electrons: lead-specific; and spectrum analyzer devices. The lead-specific type checks only for lead, while the spectrum analyzer device can be programmed to analyze for

Table 6.1. Examples of Lead Detector Devices, Manufacturers,
Measurement Ranges, and Detection Limits[a]

Lead Detection Device	Manufacturer	Measurement Range (mg/cm^2)	Detection Limit (mg/cm^2)
Spectrum analyzer MAP-3	SCITEC Corp. 200 Logton Blvd. Richland, WA 99352 (512) 251-7771	0–(no upper limit)	0.3
PGT model XK-3 lead-in-paint analyzer	Princeton Gamma-Tech, Inc. 1200 State Rd. Princeton, NJ 08540 (609) 924-7310	0.0–10	0.5
Microlead I X-ray fluorescence lead-based-paint analyzer	Warrington, Inc. 2113 Wells Branch Parkway Austin, TX 78728 (512) 251-7771	0.0–100	0.3
Spectrace 9000	Spectrace Instruments 345 East Middlefield Rd. Moutain View, CA 94043 (415) 967-0350	—	—
X-Met 880	Outokumpu Electronics 1900 N.E. Division St. #204 Bend, OR 97701 (800) 229-9209	—	—
Model SEFA-P	HNU Systems, Inc. 160 Charlemont St. Newton, MA 02161-9987 (617) 964-6690	—	—

[a]*Note:* Cost: The range of base unit costs for lead detectors is $8700–$15,690 (June 1992).

other elements as well. Various types of detectors and procedures are used to determine the quantity of lead present in a film. Table 6.1 lists some commercially available lead detection devices.

6.3. FIELD MEASUREMENT RESULTS FOR LEAD IN PAINT FILMS

6.3.1. Lead-Specific Devices

Three major field studies have been conducted to determine the precision and accuracy of measurements made using lead-specific PXRF analyzers. The results of these studies are summarized in McKnight *et al.* (1989) and

McKnight (1991). In addition, laboratory work was carried out by the National Institute of Standards and Technology (NIST) to understand further parameters thought to affect a measurement result.

A statistically designed laboratory experiment in which several parameters were randomly varied was conducted to investigate the parameters' effect on measurement precision and bias when using lead-specific analyzers. The parameters were substrate type, operator, lead concentration, including a blank, device, covering paint layer thickness, and time (to check for systematic drift). The substrate types were 12-mm thick gypsum wallboard, 10-mm-thick pine, plaster over gypsum wall board from a house built in 1940, 1-mm-thick steel, 10-mm-thick oak, 50-mm-thick clay brick, concrete block, and the pine substrate over the gypsum wallboard substrate. The lead concentrations of films made by evaporating lead on to 0.18-mm polyethylene terephthalate were 0.3, 1.2, 2.6, and 5.1 mg/cm^2. Free films of lead-free latex paint having a thickness of either 0.1 or 0.7 mm were used over the lead film to represent a configuration often found in housing with lead-containing paint films beneath nonleaded films. Two devices from each of two manufacturers, three operators, and 4 days were included in the experiment. The material and analyzer parameters were selected at random from the complete set. When dealing with steel, the variability of the blank readings over steel along the devices is of particular interest. Thus, it may be important to correct a result for a painted steel surface for substrate effect. Results from sample measurements with lead concentration of 1.2 mg/cm^2 that were corrected for substrate effect over the eight substrates were reported together with sets of data shown in McKnight *et al.* (1989).

Since the detection level of a measurement procedure is limited by the precision of a blank measurement (Currie, 1988), data were analyzed using a standard analysis of variance procedure to determine the precision of a blank measurement and the precision of lead concentration measurements near 1 mg/cm^2. The estimated precision levels for a blank observation varied from 0.20–0.39 mg/cm^2 for the four devices included in the study. For samples with lead-containing films, the precision varied from 0.26–0.49 mg/cm^2 for the four devices. The variability was primarily associated with the substrate and replicate measurements within an observation.

The procedure used in carrying out field studies to obtain estimates of precision and bias in the field measurements was the same for all three field studies. Three individual replicate PXRF measurements were made of a painted substrate, and a known area of the entire thickness of the paint film was removed for laboratory analysis. Three replicate PXRF measurements were also made of a corresponding bare substrate (*i.e.*, paint was removed by either chemical stripping or scraping). The mean of a set of results, corrected for the substrate by subtracting the mean of readings obtained on the bare substrate from that obtained on the painted substrate, was compared with results of a laboratory analysis. Laboratory analyses were carried out using the method described in ASTM D 335 (studies 1 and 2) or using a k-shell laboratory PXRF instrument (study 3) to determine total lead content.

The analysis to estimate method precision and bias was carried out by first finding the differences between the corresponding field SRF measurements and the laboratory measurement, then determining the mean and standard deviation of the difference distribution. For all three sets of field data, the estimate of the standard deviation of the difference distribution was about 0.6 mg/cm^2 for measurements made over wood, plaster, and gypsum wallboard. The means of the difference distributions were slightly greater than zero, with a pooled estimate of 0.2 mg/cm^2 for measurements over wood, plaster, and gypsum wall board. There were insufficient data points for other substrates, *i.e.*, steel, to carry out a similar analysis.

6.3.2. Spectrum-Analyzer Devices

There is only one reported set of field data that compares measurements by a spectrum analyzer device and a standard laboratory analysis procedure (McKnight *et al.*, 1989). A laboratory study was also carried out at NIST similar to the one described for lead-specific analyzers. In the laboratory study, the counting time was selected so that the precision (standard deviation) of individual replicate measurements of the same sample of gypsum wallboard was about 0.1 mg/cm^2; the results are given as follows:

Substrate	Precision (σ, mg/cm^2)
Gypsum wallboard	0.12
Wood	0.11
Steel panel	0.15
Concrete	0.24

Estimated total variances were determined for each substrate by pooling the variances over lead concentrations. The precision (square root of the total variance) reported in the preceding table is that expected for an individual reading as opposed to that for the mean of three individual readings. The analysis of variance showed that essentially all variability was due to variability of individual replicate measurements. Hence, the precision of laboratory measurements could be improved by increasing the counting time.

The same method of analysis as previously described, *i.e.*, paired comparison, was used to determine estimates of precision and bias for field measurement procedures where one individual PXRF measurement was taken from a given sample. The counting time was chosen just as it was chosen for the laboratory study. For this particular PXRF device, for lead concentrations less than 2 mg/cm^2 over wood or plaster, the standard deviation of the difference distribution is 0.3 mg/cm^2, and the mean of the difference distribution is 0.1 mg/cm^2, with two outliers excluded.

6.4. CHEMICAL SPOT TESTS

Field spot tests for lead in paint are based on a chemical reaction in which a colored lead compound is formed. Usually, these are swab tests, *i.e.*, a small amount of a test reagent(s) is placed on the film or a cut in the film and allowed to react. A reagent is selected for many properties, including its ability to react with small amounts of lead to form a visually detectable compound (sensitivity) while not reacting with other species likely to be present in paint and form a similar compound (selectivity). In addition, the chemical reaction should be tolerant of slight changes in pH, ionic strength, temperature, and other parameters that are difficult to control in a field test.

To form a colored compound in a spot test with a paint film, lead ions must be leached or separated from the lead-containing species. Dissolution of lead ions depends on the type, concentration, and dispersion of lead-containing compounds in the film and the binder, and the position of the lead-containing layer(s) in the composite film. Lead pigments used in relatively high concentrations provide hiding or control corrosion and in lower concentrations, they provide color; in still lower concentrations, they facilitate drying.

Feigl *et al.* (1972) describe several spot tests for lead, including using sodium rhodizonate, dithizone, and potassium iodide. Sulfide solutions have been used to detect lead in paint (Vind and Matthews, 1976; Sayre, 1970). Proprietary kits for detecting lead in paint based on sodium sulfide or rhodizonate are being marketed, and other kits are available for detecting lead in other species (Luk *et al.*, 1990). Results from spot test kits are discussed in the paragraphs that follow.

The paint and substrate on each test specimen were analyzed for lead using a PXRF detection unit. The PXRF unit was calibrated daily and the presence of lead was reported in units of mg/cm^2. The actual painted materials used were either flat or round. A general discussion of the measurement for lead in these materials and other shapes of painted materials follows.

Many studies have been carried out on the use of spot tests based on sodium sulfide to detect lead in paint films (McKnight *et al.*, 1989; Vind and Matthews, 1976; Sayre, 1970; Blackburn, 1990; Abramowitz, 1990). The results of these studies have been mixed. While Sayre found good correlation between sodium sulfide spot test results and lead concentration in paint films for lead concentrations greater than 0.5%, the Dewberry Davis study found a false negative rate of 29% for films having had concentrations greater than 0.7 mg/ cm^2. In the Dewberry Davis study, four operators carried out field tests on 377 paint films for which lead concentrations had been previously measured using the ASTM D 335 (1989) laboratory procedure. Also, the false negative rate did not generally decrease with increasing lead concentration.

The incidence of false positive results was not reported in the Dewberry Davis study. However, several elements sometimes found in paint films can be expected to contribute to false positive results if present in sufficient concentrations and suitable chemical form; these include copper, cobalt, mercury, manganese, and iron. Vind *et al.* (1976) reported that positive responses were not obtained under typical testing conditions for iron oxide pigments. In contrast, they found that neat biocides (containing mercury or copper) and driers (containing cobalt or manganese) did turn black in the presence of sulfide ions. However, they concluded that given the usually low concentrations of these materials in paints, interference due to these elements would be minimal.

Preliminary laboratory studies to evaluate the characteristics of proprietary spot test kits based on sodium rhodizonate have been completed by Lin-Fu (1985). These studies showed that a color change indicative of the presence of lead occurred in laboratory-prepared paint films at an area concentration of about 1 mg/cm^2; a color change was not observed for films with lead concentrations less than 0.1 mg/cm^2. Studies also reported that these kits could detect about 0.1 μg of lead in a water-soluble salt. This implies that lead ions from the lead compound in the paint films were not being readily leached by the solution used in the kit. Further, it was reported that detection level was lower for lead nitrate than lead chloride.

Another study, sponsored by HUD, on the use of sodium sulfide to detect lead in paint is being carried out at NIST. The objectives of this study include: (1) investigating additional candidate procedures for carrying out a field chemical test and (2) determining estimates of the precision and accuracy of a potential candidate alternative procedure and a swab spot test procedure, when carried out by trained lead-based paint inspectors.

6.5. MEASUREMENT ACCURACY CONCLUSIONS

Two general types of field measurements for detecting lead in paint, PXRF and chemical spot tests are available. Although it is desirable to have a field method that is nondestructive, rapid, easy to perform, nonbiased, precise, safe, and inexpensive, both field procedures discussed have some limitations.

6.5.1. PXRF Test Limitations

Measurements based on PXRF devices are affected by substrate underlying the paint film. For the lead-specific devices, it is necessary to correct a paint film reading for substrate effect, thus making it a somewhat destructive measurement. With a substrate correction, we can expect that the true lead concentration should be within $\pm$ 1.4 mg/cm^2 of the experimental outcome 95% of the time for measurements over wood, gypsum wallboard, and plaster.

6.5.2. Spectrum Analyzer Test Limitations

Based on data for the spectrum analyzer device, a substrate correction may not be needed for some substrates, but based on laboratory data, it may be needed for other substrates, such as steel. From the limited data available, the true lead concentration is expected to be within ± 0.7 mg/cm^2 of the experimental value 95% of the time for measurements over wood and plaster when the precision of replicate measurements over the same substrate is no poorer than 0.1 mg/cm^2.

6.5.3. Correcting Irregular Geometric Shapes with PXRF Devices

Irregular shapes of house components have not formerly been addressed, but it is recommended that measurement of any irregular shape, such as a round post, molding, *etc.*, be especially addressed, since the flat-shaped measurement will obviously produce a different measurement. Measuring irregularly shaped wood and metal surfaces is discussed by Gooch (1991) in a Georgia Tech Research Institute report on lead paint abatement.

6.5.4. Chemical Spot Test Limitations

There are chemicals that can be used as indicators that are sensitive to low concentrations of lead and also reasonably selective. Based on field experience, there is considerable confidence in a sodium sulfide spot test procedure for testing for lead in paint. On completing further studies being supported by HUD and the EPA, more definitive statements about chemical spot field tests should be possible.

6.6. DETECTING LEAD IN AIR

The working environment and atmosphere can be analyzed for lead by filtering air close to a worker's head, then analyzing the collected material on the filter for lead. Lead is usually reported in mg/m^3. The basic procedures used for monitoring air are taken from "OSHA Instruction CPL 2-2.20B CH-1," November 13, 1990, Directorate of Technical Support, pp. 1-1, 1-2, 1-3.

6.6.1. Collecting Lead from Air

The procedure for sampling the air in the work environment consists of (1) placing a polyvinyl chloride (PVC) tube within 12 inches of the operator's mouth; (2) blowing air from the breathing area at 2 liter per minute into the tube and through a 0.45 μ filter supplied in a two-piece cassette; and (3) collecting lead paint particulate for 8 hours. A portable pump was used to blow the air at negative pressure through the filter from the breathing area.* These materials are listed below.

6.6.2. Analyzing Collected Lead

Lead particulate collected on each filter was analyzed by atomic emission spectroscopy. The procedure included digesting the filter in concentrated nitric acid to dissolve lead compounds. The solution of lead was injected in a Perkin–Elmer Atomic Emission Spectrophotometer model no. 2380. The total mass of lead from each filter was correlated to the total air volume and reported in mg/m^3.

---

*The following products are used to collect lead samples from the air: MSA Flow-Lite Pro portable pump, model 484107; Wisconsin Occupational Health Laboratory 0.45 μ; Cole–Parmer Corporation PVC tubing.

7

Laboratory Analysis Methods

The following laboratory methods and techniques for analyzing paint have been used successfully by the author. Methods of analysis are listed by application and specific property in Table 7.1. Each method is discussed in the following paragraphs.

- Bulk analysis of lead in paint
    Atomic spectroscopy (AS)
    X-ray diffraction
- Microanalysis of lead in paint
    Electron beam probe (EBP)
    Electron spectroscopy for chemical analysis
    Light microscopy
    Electron microscopy
- Analysis of binders in paint
    Infrared spectroscopy
    Electron spectroscopy for chemical analysis

7.1. ATOMIC SPECTROSCOPY

This is actually not one technique but three (Willard *et al.*, 1974): atomic absorption, atomic emission, and atomic fluorescence. Of these, atomic absorption (AA) and atomic emission are the most widely used. Out discussion focuses on them and an affiliated technique, ICP-Mass Spectrometry.

Table 7.1. Properties and Preferred Methods of Determination

Method	Cross Sections and Topological Sections	Crystalline Structure	Elements Present	Particle Size	Binder Identification
			Properties		
Atomic spectroscopy	No		1, 2[a]		
X-ray diffraction	Yes	1, 2			
Electron beam probe	Yes		1, 2	2	
Electron spectroscopy for chemical analysis	Yes		1	1	1
Light microscopy	Yes			2	
Electron microscopy (with EDXRA)	Yes		1	2	
Infrared spectroscopy	No				1

Note: [a] 1 = Qualitative, 2 = Quantitative.

7.1.1. Atomic Absorption

This process occurs when a ground state atom absorbs energy in the form of light of a specific wavelength and is thereby elevated to an excited state (Willard *et al.*, 1974). The amount of light energy absorbed at this wavelength increases as the number of atoms of the selected element in the light path increases. The relationship between the amount of light absorbed and the concentration of analyte present in known standards can be used to determine unknown concentrations by measuring the amount of light they absorb. Instrument readouts can be calibrated to display concentrations directly.

The basic instrumentation for AA requires a primary light source, an atom source, a monochromator to isolate specific wavelength of light used, a detector to measure light accurately, electronics to treat the signal, and a data display or logging device to show the results. The atom source used must produce free analyte atoms from the sample. The source of energy for free atom production is heat, most commonly in the form of an air–acetylene or nitrous oxide–acetylene flame. The sample is introduced as an aerosol into the flame. The flame burner head is aligned so that the light beam passes through the fame, where the light is absorbed.

7.1.2. Graphite Furnace Atomic Absorption

The major limitation of AA using flame sampling (flame AA) is that the burner–nebulizer system is a relatively inefficient sampling device. Only a small fraction of the sample reaches the flame, and the atomized sample passes quickly through the light path. An improved sampling device atomizes the entire sample and retains the atomized sample in the light path for an extended period to enhance the sensitivity of the technique. Electrothermal vaporation using a graphite furnace provides these features.

With graphite furnace atomic absorption (GFAA), the flame is replaced by an electrically heated graphite tube. The sample is introduced directly into the tube, which is then heated in a programmed series of steps to remove the solvent and major matrix components, then atomize the remaining sample. All of the analyte is atomized, and the atoms are retained within the tube (and the light path, which passes through the tube) for an extended period. As a result, sensitivity and detection limits are significantly improved.

Graphite furnace analysis times are longer than those for flame sampling, and fewer elements can be determined using GFAA. However, the enhanced sensitivity of GFAA and the ability of GFAA to analyze very small samples and directly analyze certain types of solid samples significantly expand the capabilities of AA.

7.1.3. Atomic Emission

Atomic emission spectroscopy (Willard *et al.*, 1976; Dean *et al.*, 1974) is a process where light emitted by excited atoms or ions is measured. The emission occurs when sufficient thermal or electrical energy is available to excite a free atom or ion to an unstable energy state. Light is emitted when the atom or ion returns to a more stable configuration or the ground state. The wavelengths of light emitted are specific to the elements present in the sample.

The basic instrument used for atomic emission is very similar to that used for AA with the difference that no primary light source is used for atomic emission. One of the more critical components for atomic emission instruments is the atomization source (Grove, 1971) because it must also provide sufficient energy to excite the atoms as well as atomize them.

The earliest energy sources for excitation were simple flames, but these often lacked sufficient thermal energy to be truly effective. Later, electrothermal sources, such as arc/spark systems were used, particularly when analyzing solid samples. These sources are useful for qualitative and quantitative work with solid samples, but they are expensive, difficult to use, and have limited applications.

Due to the limitations of early sources, atomic emission did not initially enjoy the universal popularity of AA. This changed dramatically with the development of the inductively coupled plasma (ICP) as a source for atomic emission. The ICP eliminates many of the problems associated with past emission sources and has caused a dramatic increase in the utility and use of emission spectroscopy.

7.1.4. Inductively Coupled Plasma

The ICP (Bertin, 1970) is an argon plasma maintained by the interaction of an radio frequency (RF) field and ionized argon gas. The ICP is reported to reach temperatures as high as 10,000 K, with the sample experiencing useful temperatures between 5500–8000 K. These temperatures allow complete atomization of elements, minimizing chemical interference effects.

Plasma is formed by a tangential stream of argon gas flowing between two quartz tubes. The RF power is applied through the coil, and an oscillating magnetic field is formed. Plasma is created when the argon is made conductive by exposing it to an electrical discharge that creates seed electrons and ions. Inside the induced magnetic field, the charged particles (electrons and ions) are forced to flow in a closed annular path. As they meet resistance to their flow, heating takes place and additional ionization occurs. The process occurs almost instantaneously, and the plasma expands to its full dimensions.

As viewed from the top, the plasma has a circular, doughnut shape. The sample is injected as an aerosol through the center of the doughnut. This characteristic of the ICP confines the sample to a narrow region and provides an optically thin emission source and a chemically inert atmosphere. This results in a wide dynamic range and minimal chemical interactions in an analysis. Argon is also used as a carrier gas for the sample.

7.1.5. ICP-Mass Spectroscopy

As its name implies, ICP-mass spectrometry (ICP-MS) is the synergistic combination of an inductively coupled plasma with a quadrupole mass spectrometer (Birks, 1959). The ICP-MS uses the ability of the argon ICP to generate efficiently singly charged ions from the elemental species within a sample. These ions are then directed into a quadrupole mass spectrometer.

The function of the mass spectrometer is similar to that of the monochromator in an AA or ICP emission system. However, rather than separating light according to its wavelength, the mass spectrometer separates the ions introduced from the ICP according to their mass to charge ratio. Ions of the selected mass/charge are directed to a detector that quantitates the number of ions present. Due to the similarity of the sample introduction and data handling techniques, using an ICP-MS is very much like using an ICP emission spectrometer.

The ICP-MS combines the multielement capabilities and broad linear working range of ICP emission with the exceptional detection limits of GFAA. It is also one of the few analytic techniques that permit quantifying elemental isotopic concentrations and ratios.

7.1.5.1. Selecting the Proper Atomic Spectroscopy Technique

With the availability of a variety of atomic spectroscopy techniques, such as flame AA, GFAA, ICP emission, and ICP-MS, laboratory managers must decide which technique is best suited for their laboratory's analytic problems. Important criteria for selecting an analytic technique include detection limits, analytic working range, sample throughput, cost, interferences, ease of use, and availability of proven methodology. These criteria are discussed in the following paragraphs for flame AA, GFAA, ICP emission, and ICP-MS.

7.1.5.2. Atomic Spectroscopy Detection Limits

Achievable detection limits for individual elements represent a significant criterion of the usefulness of an analytic technique for a given analytic problem. Without adequate detection limit capabilities, lengthy analytic concentration procedures may be required prior to analysis.

Typical detection limit ranges for the major atomic spectroscopy techniques are listed in Table 7.2 which provides detection limits of element for five atomic spectroscopic techniques: flame AA, mercury/hydride AA, GFAA, ICP emission, and ICP-MS. Interferences for these methods are listed in Table 7.3.

Generally, the best detection limits are attained using ICP-MS or GFAA. For mercury and those elements that form hydrides, the cold vapor mercury or hydride generation techniques offer exceptional detection limits. Most manufacturers (*e.g.*, Perkin–Elmer) define detection limits very conservatively with either a 95% or 98% confidence level, depending on established conventions for the analytic technique. This means that if a concentration at the detection limit were measured many times, it would be distinguished from a zero or baseline reading in 95% (or 98%) of the determinations.

7.2. X-RAY DIFFRACTION

Every atom in a crystal scatters an X-ray beam (Bertin, 1970) incident on it in all directions. Since even the smallest crystal contains a very large number of atoms, the chance that these scattered waves will constructively interfere is almost zero except for the fact that atoms in crystals are arranged in a regular, repetitive manner. The condition for diffraction of an X-ray beam from a crystal is given by the Bragg equation (Birks, 1959, 1963; Bunn, 1961, Clark, 1955). Atoms located exactly on crystal planes contribute maximally to the intensity of the diffracted beam; atoms exactly halfway between the planes exert maximum destructive interference; and those at some intermediate location interfere constructively or destructively, depending on their exact location, but with less than their maximum effect. Furthermore, the scattering power of an atom for X-rays depends on the number of electrons it possesses. Thus, the position of the diffraction beams from a crystal depends only on the size and shape of the repetitive unit of a crystal and the wavelength of the incident X-ray beam, whereas intensities of the diffracted beams also depend on the type of atoms in the crystal and the location of the atoms in the fundamental repetitive unit, the unit cell (Henke *et al.*, 1970, Liebhafsky *et al.*, 1960; Liebhafsky *et al.*, 1964). No two substances, therefore, have absolutely identical diffraction patterns when we consider both the direction and intensity of all diffracted beams (Robertson, 1953; Sproull, 1946); however, some similar, complex organic compounds may have almost identical patterns. The diffraction pattern is thus a fingerprint of a crystalline compound, and crystalline components of a mixture can be identified individually.

Table 7.2. Atomic Spectroscopy Detection Limits (µg/l)

Element	Flame AA	Hg/Hydride	GFAA	ICP Emission	ICP-MS
Ag	0.9		0.005	1	0.003
Al	30		0.04	4	0.006
As	100	0.02	0.2	20	0.006
Au	6		0.1	4	0.001
B	700		20	2	0.09
Ba	8		0.1	0.1	0.002
Be	1		0.01	0.06	0.03
Bi	20	0.02	0.1	20	0.0005
Br					0.2
C				50	150
Ca	1		0.05	0.08	2
Cd	0.5		0.003	1	0.003
Ce				10	0.0004
Cl					10
Co	6		0.01	2	0.0009
Cr	2		0.01	2	0.02
Cs	8		0.05		0.0005
Cu	1		0.02	0.9	0.003
Dy	50				0.001
Er	40				0.0008
Eu	20				0.0007
F					10000
Fe	3		0.02	1	0.4
Ga	50		0.1	10	0.001
Gd	1200				0.002
Ge	200		0.2	10	0.003
Hf	200				0.0006
Hg	200	0.006	1	20	0.004
Ho	40				<0.0005
I					0.008
In	20		0.05	30	<0.0005
Ir	600		2	20	0.0006
K	2		0.02	50	1
La	2000			1	0.0005
Li	0.5		0.05	0.9	0.03
Lu	700				<0.0005
Mg	0.1		0.004	0.08	0.007
Mn	1		0.01	0.4	0.002
Mo	30		0.04	5	0.003
Na	0.2		0.05	4	0.05
Nb	1000			3	0.0009
Nd	1000				0.002
Ni	4		0.1	4	0.005
Os	80				

(Continued)

Table 7.2.   (*Continued*)

Element	Flame AA	Hg/Hydride	GFAA	ICP Emission	ICP-MS
P	50000		30	30	0.3
Pb	10		0.05	20	0.001
Pd	20		0.25	1	0.003
Pr	5000				<0.0005
Pt	40		0.5	20	0.002
Rb	2		0.05		0.003
Re	500			20	0.0006
Rh	4			20	0.008
Ru	70			4	0.002
S				50	70
Sb	30	0.1	0.2	60	0.001
Sc	20			0.2	0.02
Se	70	0.02	0.2	60	0.06
Si	60		0.4	3	0.7
Sm	2000				0.001
Sn	100		0.2	40	0.002
Sr	2		0.02	0.05	0.0008
Ta	1000			20	0.0006
Tb	600				<0.0005
Te	20	0.02	0.1	50	0.01
Th					<0.0005
Ti	50		1	0.5	0.006
Tl	9		0.1	40	0.0005
Tm	10				<0.0005
U	10000			10	<0.0005
V	40		0.2	2	0.002
W	1000			20	0.001
Y	50			0.2	0.0009
Yb	5				0.001
Zn	0.8		0.01	1	0.003
Zr	300			0.8	0.004

Table 7.3.   Atomic Spectroscopy Interferences

Technique	Type of Interference	Method of Compensation
Flame AA	Ionization	Ionization buffer
	Chemical	Releasing agent or nitrous oxide-acetylene flame
	Physical	Dilution, matrix matching, or method of additions
GFAA	Physical and chemical molecular absorption	STPF conditions Zeeman or continuum source background correction
	Spectral	Zeeman background correction
ICP emission	Spectral	Background correction or the use of analytic lines
ICP-MS	Mass overlap	Interelement correction, alternative mass values, or higher mass resolution

7.2.1. Reciprocal Lattice Concept

Diffraction phenomena can be interpreted most conveniently with the aid of the reciprocal lattice concept. A plane can be represented by a line drawn normal to the plane; the spatial orientation of this line describes the orientation of the plane. Furthermore, the length of the line can be fixed in an inverse proportion to the interplanar spacing of the plane it represents.

When a normal is drawn to each plane in a crystal and normals are drawn from a common origin, the terminal points of these normals constitute a lattice array. This is called the reciprocal lattice (Birks, 1953; Bragg, 1933) because of the distance of each point from the origin is reciprocal to the interplanar spacing of the planes that it represents. In an individual cell of a crystalline structure, there exists near the origin the traces of several planes in a unit cell of a crystal, namely, the (100), (001), (101), and (102) planes. Normals to these planes are called the reciprocal lattice vectors α_{hkl}, and they are defined by

$$\alpha_{hkl} = \frac{\lambda}{d_{hkl}}$$

In three dimensions, the lattice array is described by three reciprocal lattice vectors whose magnitudes are given by

$$a^* = \alpha_{100} = \frac{\lambda}{d_{100}}$$

$$b^* = \alpha_{101} = \frac{\lambda}{d_{101}}$$

$$c^* = \frac{\alpha}{001} = \frac{\lambda}{d_{001}}$$

There directions are defined by three interaxial angles α^*, β^*, γ^*.

Writing the Bragg equation in a form that relates the glancing angle θ most clearly to the other parameters, we have

$$\sin \theta_{hkl} = \frac{\lambda/d_{khkl}}{2}$$

The numerator can be taken as one side of a right triangle, with θ as another angle and the denominator as its hypotenuse. The diameter of a circle represents the direction of the incident X-ray beam. A line through the origin of the circle and forming the angle θ with the incident beam represents a crystallographic plane that satisfies the Bragg diffraction condition. A line forming the angle θ with the crystal plane and 2θ with the incident beam represents the diffracted beam's direction. Another line is the reciprocal lattice vector to the reciprocal lattice point P_{hkl} lying on the circumference of the circle. The vector

α_{hkl} originates at the point on the circle where the direct beam leaves the circle. The Bragg equation is satisfied when and only when a reciprocal lattice points lies on the sphere of reflection, that is, a sphere formed by rotating the circle on the diameter.

Thus, the crystal in a diffraction experiment can be pictured at the center of a sphere of unit radius, and the reciprocal lattice of this crystal is centered at the point where the direct beam leaves the sphere. Because the orientation of the reciprocal lattice bears a fixed relation to that of the crystal, the reciprocal lattice can also be pictured as rotating if the crystal is rotated. When a reciprocal lattice point intersects the sphere, a reflection emanates from the crystal at the sphere's center and passes through the intersecting reciprocal lattice point.

7.2.1.1. Diffraction Patterns

If the X-ray beam is monochromatic, there will be only a limited number of angles where diffraction of the beam can occur. The actual angles are determined by the wavelength of the X-rays and spacing between various crystal planes. In the *rotating crystal method*, monochromatic X radiation is incident on a single crystal rotated around one of its axes. The reflected beams lie as spots on the surface of cones coaxial with the rotation axis.

Diffracted beam directions are determined by intersecting the reciprocal lattice points with the sphere of reflection. All the reciprocal lattice points lying in any one layer of the reciprocal lattice layer perpendicular to the axis of rotation intersect the sphere of the reflection in a circle. The height of the circle above the equatorial plane is proportional to the vertical reciprocal lattice spacing. By remounting the crystal successively around different axes, we can determine the complete distribution of reciprocal lattice points. Of course, one mounting is sufficient if the crystal is cubic, but two or more may be needed if the crystal has lower symmetry.

In a modification of the single-crystal method, known as the Weissenberg method, a photographic film is continuously moved during exposure parallel to the axis of rotation of the crystal. All reflections are blocked out except those that occur in a single layer line. This results in a film that is somewhat easier to decipher than a simple rotation photograph. Still other techniques are used: The *precession method* produces a photograph with an undistorted view of a plane in the reciprocal lattice of the crystal.

In the *powder method*, the crystal is replaced by a large collection of very small randomly oriented crystals and a continuous cone of diffracted rays is produced. There are some important differences, however, with respect to the rotating-crystal method. The cones obtained with a single crystal are not continuous because diffracted beams occur only at certain points along the cone, whereas cones obtained by the powder method are continuous. Furthermore, although the cones obtained from rotating single crystals are uniformly spaced around the zero level, cones produced by the powder method are determined by spacings in prominent planes, and therefore these cones are not uniformly spaced. Because of the random orientation of the crystallites, the reciprocal lattice points generate a sphere of radius α_{hkl} around the origin of the reciprocal lattice. A number of these spheres intersect the sphere of reflection.

7.2.1.2. Camera Design

Cameras are usually constructed so that the film diameter has one of three values, 57.3 mm, 114.6 mm, or 143.2 mm. The reason for these values can be understood by considering the calculations involved. If the distance between corresponding arcs of the same cone of diffracted rays is measured and called S, then

$$4\theta_{rad} = \frac{S}{R}$$

where θ_{rad} is the Bragg angle measured in radians and R is the radius of the film in the camera. The angle θ_{deg}, measured in degrees, is then

$$\theta_{deg} = \frac{57.295S}{4R}$$

where 57.295 equals the value of a radian in degrees. Therefore, when the camera diameter ($2R$) is equal to 57.3 mm, $2\theta_{deg}$ may be found by measuring S in millimeters. When the diameter is 114.59 mm, $2\theta_{deg} = S/2$; when the diameter is 143.2 mm, $\theta_{deg} = 2(S/10)$.

Once angle θ has been calculated, the preceding equation can be used to find the interplanar spacing, using values of wavelength λ. Tables that give the interplanar spacing for the angle 2θ for the types of radiation most commonly used are available.

Cameras of larger diameter make it easier to measure the separation of lines if the lines are sharp. The sharpness of the lines depends to a large extent on the quality of the collimating slits and the size of the sample. The slits should produce a fine beam of X-rays with as small a divergence as possible. The sample size should be small so that it will act as a small source of the diffracted beam. On the other hand, smaller samples, finer pencils of incident X-rays, and the cameras of larger diameter all tend to require longer exposure times, so that, in practice, a compromise must be made.

For very precise measurements of interplanar spacings, the diameter of the film and the separation of the lines must be very accurately known. Several measurement methods have been proposed. The effective camera diameter can be determined by calibration with a material, such as sodium chloride, whose interplanar spacings are accurately known. In the Straumanis method, film is inserted in the camera so that the ends of the film are at about 90° from the point of emergence of the beam from the camera. If a_1, a_2, and b_1, b_2 represent the two sides of arcs on the left and right sides of the film, then the two averages determine the positions of the entering and emerging beams. Therefore,

$$\frac{b_1 + b_2}{2} - \frac{a_1 + a_2}{2} = \text{distance corresponding to } 180°$$

or

$$360° = b_1 + b_2 - a_1 - a_2$$

The angle 4θ associated with any pair of lines can then be calculated

$$\frac{a_2 + a_1}{b_1 + b_2 - a_1 - a_2} = \frac{4\theta}{360}$$

7.2.1.3. Choice of X Radiation

Two factors control the choice of X radiation, as can be seen by rearranging the terms of the Bragg equation

$$\theta = \sin^{-1}\left(\frac{\lambda}{2d}\right)$$

Because the ratio in the parentheses cannot exceed unity, the use of long-wavelength radiation limits the number of reflections that can be observed. Conversely, when the unit cell is very large, short-wavelength radiation tends to crowd individual reflections very close together.

The choice of radiation is also affected by the absorption characteristics of a sample. Radiation having a wavelength just shorter than the absorption edge of an element contained in the sample should be avoided, because then the element absorbs radiation strongly. Absorbed energy is emitted as fluorescent radiation in all directions and increases the background (which would result in darkening on a film, making it more difficult to see the diffraction maxima sought). It is obvious, then, why one commercial source provides radiation sources from a multiwindow tube with anodes of silver, molybdenum, tungsten, or copper.

7.2.1.4. Samples

Single crystals are used for structure determinations when possible because of the relatively larger number of reflections obtained from single crystals and the greater ease of their interpretation. Crystal size should be such that it is completely bathed by the incident beam. Generally, a crystal is affixed to a thin glass capillary that is in turn fastened to a brass pin.

When single crystals of sufficient size are not available or when the problem is merely to identify a material, a polycrystalline aggregate is formed into a cylinder whose diameter is smaller than the diameter of the incident X-ray beam. Metal samples are machined to a desirable shape, plastic materials can often be extruded through suitable dies, and all other samples are best ground to a fine powder (200–300 mesh) and shaped into thin rods after mixture with a binder (usually collodion).

Although liquids cannot be identified directly, it is frequently possible to convert them into crystalline derivatives that have characteristic patterns. Many of the classical derivatives can be used to identify aldehydes and ketones as 2,4-dinitrophenol-hydrazones, fatty acids as p-bromoanilides, and amines as picrate derivatives.

To obtain diffraction patterns from large, dense samples, a back-reflection camera can be used. The X-rays pass through an opening in the center of the film and impinge on the sample. Beams diffracted over a range of Bragg angles extending from 59–88° are registered on the circular film.

7.2.1.5. Automatic Diffractometers

Results are achieved rapidly and with much better precision when automatic diffractometers are used to record diffraction data. The X-ray tube furnishes the radiation directly (filters are generally used to obtain more nearly monochromatic radiation). The diffracting crystal is replaced by a powdered or metallic sample. To increase the randomness of orientation in the crystallites, the sample can be rotated in its own plane, that is, the plane perpendicular to the bisector of the angle between the source and detector beams. Note also that as the sample is rotated in the other plane to sweep through various θ angles, the detector must be rotated twice as rapidly to maintain the angle 2θ with the irradiating beam.

When used as detectors, scintillation and proportional counters, with their associated circuitry, are far superior to photographic film in regard to the number of reflections per day that can be recorded. They can achieve a precision of 1% or better. Even with the best darkroom and photometric procedures, the relative degree of blackness of each spot or cone on the film cannot be estimated with an accuracy of much more than 10%, and often the error in estimation is greater than this. The principal advantage of photographic film over counters is that it provides a means of recording many reflections at one time.

Automatic single-crystal diffractometers are quite complex. A cradle assembly provides a wide angular range for orienting and aligning the crystal under study. A precision diffractometer assembly allows the detector to trans-verse a spherical surface from longitude $-5°$–$150°$ and from latitude $-6°$–$60°$ on the Norelco instrument. The complete unit provides four rotational degrees of freedom for the crystal and two for the detector. Various crystal and counter angles are set on the basis of programmed information, and the resulting diffraction intensity is measured as a function of angle. Single-crystal diffrac-tometers are being widely used to gather data necessary for crystal structure determinations.

7.2.1.6. X-Ray Powder Data File

For paint purposes, to identify a powder (pigments) sample is desired, its diffraction pattern is compared with diagrams of known substances until a match is obtained. This method requires a library of standard films to be available. Alternatively, d values calculated from the diffraction diagram of the unknown substance are compared with the d values of over 5000 entries listed on plain cards, Keysort cards, and IBM cards in the X-ray powder data file (Switzer *et al.*, 1948). An index volume is available with the file. The cataloging scheme (American Society of Testing Materials, 1989) used to classify different cards lists the three most intense reflections in the upper left corner of each card. The cards are then arranged in sequence of decreasing d values of the most intense reflections, based on 100 for the most intense reflection observed.

To use the file to identify a sample containing one component, first look for the d value of the darkest line of the unknown in the index. Since there is probably more than one listing for the first d value, d values of the next two darkest lines are then matched against the values listed. Finally, the various cards involved are compared. A correct match requires all the lines on the card and film to agree. It is also good practice to derive the unit cell from the observed interplanar spacings and to compare it with that listed on the card.

If the unknown contains a mixture, each component must be identified individually. This is done by treating the list of d values as if they belonged to a single component. After a suitable match for one component is obtained, all the lines of the identified component are omitted from further consideration. The intensities of the remaining lines are rescaled by setting the strongest intensity equal to 100 and repeating the entire procedure.

Reexamining the cards in the file is a continuing process to eliminate errors and remove deficiencies. Replacement cards for substances bear a star in the upper right corner.

X-ray diffraction furnishes a rapid, accurate method for identifying crystalline phases present in a material. Sometimes, it is the only method available for determining which of the possible polymorphic forms of a substance are present—for example, carbon in graphite or diamond. Differentiating among various oxides, such as FeO, Fe_2O_3, and Fe_3O_4 or between materials present in such mixtures as $KBr + NaCl$, $KCl + NaBr$, or all four, is easily accomplished with X-ray diffraction, whereas chemical analysis indicates only the ions present, and not the actual state of combination. The presence of various hydrates is another possibility.

Appendix A contains a collection of X-ray powder data files for lead pigments.

7.2.1.7. Quantitative Analysis

X-ray diffraction is adaptable to quantitative applications because the intensities of the diffraction peaks of a given compound in a mixture are proportional to the fraction of the material in the mixture. However, directly comparing the intensity of a diffraction peak in the pattern obtained from a mixture is difficult. Corrections are frequently necessary for differences in absorption coefficients between the compound being determined and the matrix. Preferred orientations must be avoided. Internal standards help, but they do not overcome the difficulties entirely.

7.2.1.8. Structural Applications

A discussion of the complete structural determination for a crystalline substance is beyond the scope of this book. Ample references are given to pertinent literature, and it suffices to point out that with careful work, atoms can be located to a precision of hundredths of an angstrom unit or better.

7.2.1.9. Crystal Topography

There are a number of experimental diffraction techniques developed in recent years by which the microscopical defects in a crystal can be shown. Most crystals are far from perfect and exhibit regions (grains) with somewhat differing orientations, or they may contain such individual defects as dislocations or faults distributed throughout the crystal. Studies of these defects are important in understanding the nature of stress in metals, the nature and behavior of "doped" crystals used in transistors, the production of perfect crystals, and other phenomena.

Microradiographic methods are based on absorption, and contrast in the images is due to differences in absorption coefficients from point to point. X-ray diffraction (XRD) topography depends for image contrast on point-to-point changes in the direction or intensity of beams diffracted by planes in the crystal.

One much-used method of X-ray diffraction topography is the Berg–Barrett method, where the crystal is set to reflect X-rays at the Bragg angle for some plane. Geometric resolutions of about 1 μm can be achieved, and single dislocations can be resolved. The contrast on the film is due to variations in the reflecting power due to imperfections in the crystal.

Another method of X-ray diffraction topography is the Lang method, where a ribbon X-ray beam is collimated to such a small angular divergence that only one characteristic wavelength is diffracted by the crystal. Simultaneous movement of the crystal and film allow a large area of the crystal to be investigated.

7.3. ELECTRON BEAM PROBE

Electron beam probe microanalysis, developed by Castaing (1951), is a method of the nondestructive elemental analysis from an area only 1 μm in diameter at the surface of a solid specimen. A beam of electrons is collimated into a fine pencil of 1-μm cross section and directed at a specimen surface exactly on the spot to be analyzed. This electron bombardment excites characteristic X-rays essentially from a point source and at intensities considerably higher than with fluorescent excitation. The limit of detectability (in a 1-μm sized region) is about 10^{-14} g. The relative accuracy is 1–2% if the concentration is greater than a few percentages and adequate standards are available.

Three types of optics are employed in the microprobe spectrometer: electron optics, light optics, and X-ray optics. Of these, the most complex is the electron optical system, a modified electron microscope consisting of an electron gun and two electromagnetic focusing lenses to form the EBP. The specimen is mounted as the target inside the vacuum column of the instrument and under the beam. A focusing, curved crystal X-ray spectrometer is attached to the evacuated system with the focal spot of the electron beam serving as the source of X radiation. A viewing microscope and mirror system allow continuous visual observation of the exact area of the specimen that the electron beam is striking. Point-by-point microanalysis is accomplished by translating the specimen across the beam. This method (Wittry, 1964) is used to study variations in concentration occurring near gain boundaries, to analyze small inclusions in alloys or precipitates in a multitude of products, and in corrosion studies, where excitation is restricted to thin surface layers because the beam penetrates only 1 or 2 μm into the specimen.

7.4. ELECTRON SPECTROSCOPY FOR CHEMICAL ANALYSIS

When X-rays impinge on matter, electrons may be ejected, as described previously. If monoenergetic X-rays are used, the kinetic energy of the ejected electrons E_k is determined by the difference between the binding energy of the electron E_b and the energy of the X-ray photon $h\nu$. If the kinetic energy of the ejected electrons is determined, the binding energy of the electrons can be evaluated by the relationship

$$E_b = h\nu - E_k$$

Since the binding energies of electrons in the outer region of the atomic core depend on the chemical environment, this method, described by Siegbahn (1967) and Siegbahn *et al.* (1969), can be used to analyze several types of mixtures that would be difficult by other means. For example, Swartz *et al.* (1971) have developed a quantitative measurement for MoO_2 and MoO_3 mixtures. There is a 1.7-eV increase in the binding energy of the $3d_{3/2}$ and $3d_{5/2}$ electrons ejected from the Mo(VI) oxide compared to the Mo(IV) oxide. The $3d_{3/2}$ electrons have a binding energy about 3.1 eV higher than do $3d_{5/2}$ electrons.

In practice, samples are irradiated by monoenergetic X-rays, and emitted electrons are analyzed by passing them through a double-focusing electron spectrometer similar to those used to study beta ray energies. Swartz *et al.* (1971) used the aluminum $K\alpha_1$, X-ray line at 1487 eV, for their work on molybdenum oxides, a Bendix Channeltron Electron Multiplier detector was used to count the electrons. For the molybdenum oxide analyses, synthetic samples were used to obtain the calibration curve relating concentration of MoO_2 to the ratio of counts of $Mo(3d_{32})$ electrons in MoO_3 to $Mo(3d_{52})$ electrons in MoO_2.

Electron spectroscopy for chemical analysis (ESCA) is primarily suitable for surface analyses because ejected electrons are easily stopped by even a minute thickness of solid. Thus, the analysis is characteristic of the few top monolayers of a surface. This means that a chip of paint can be fractured to expose the cross-sectional view to allow pigments and binders to be analyzed. All elements except hydrogen can be identified, and the oxidation state and bonding of the element are usually determinable. The ESCA has been used to study changes responsible for catalysts poisoning, surface contaminants on semiconductors that may cause poor performance, surface reactions on metals; to characterize textile fiber surfaces, and many other surface phenomena, including human skin composition.

7.5. LIGHT MICROSCOPY

Light microscopy (LM) (Kane and Larrabee, 1974) is useful for studying pigments for color, particle size and distribution, and concentration in films. Although LM can be used to study paint surfaces as well, scanning electron microscopes (SEMs) have taken over that function. The SEM resolves detail one-tenth as large (20 nm = 0.02 μm) as that resolved by the light microscope, and the in-focus depth of field for the SEM is 100–300 times that of the light microscope.

There are other advantages to the SEM, including ease of sample preparation, elemental analysis by energy-dispersive X-ray analyzer, and usually excellent specimen contrast. The light microscope is still important because an SEM costs 10–50 times the cost of an adequate light microscope. Secondly, there are many routine surface examinations easily performed by light optics that do not justify use of the SEM.

There are at least a few surface characterization problems that the SEM cannot do: surfaces of materials unstable under high-vacuum or high-energy electron bombardment; samples too bulky for the SEM sample compartment; and finally, samples requiring manipulation on the surface during examination and vertical resolution of detail below 250 μm. Also, the natural color of the specimen (*e.g.*, paint pigments) can be observed with the light microscope, while it cannot by an SEM.

Often of course both the LM and the SEM are used to examine paint materials. The stereobionocular microscope, an LM, is needed if only to quickly decide what areas to study or to examine pertinent areas in terms of the total sample, including color. Even SEM examination should begin at low magnification and never been increased more than necessary.

Accessories greatly enhance the light microscope's ability to resolve detail, differentiate between compositions, or increase contrast. Any microscopist who has attempted to observe thin coatings on paper, *e.g.*, ink lines, with the SEM soon returns to the light microscope. The Nomarski interference contrast system on a reflected light microscope gives excellent rendition of surface detail for metals, ceramics, polymers, or biological tissue. The SEM is 10 times better than the light microscope in horizontal resolution but 20 times worse in vertical resolution.

To microscopists, surface characterization means topography, elemental composition, and solid-state structure. They usually study all three by what is usually termed morphological analysis, *i.e.*, shape characteristics. Surface geometry or topography is obviously a function of morphology, although the light microscopist may have to enhance contrast of transparent, colorless surfaces like paper or ceramics with a surface treatment, *e.g.*, an evaporated metal coating or by a special examination technique, *e.g.*, interferometry.

It is also often possible to determine elemental composition by studying morphology, although it is again perhaps made easier by surface etching, staining, or polarized light examination. Finally, solid-state structure is also often apparent morphologically, *e.g.*, the various phases in steels (austenite, ferrite, pearlite, sorbite, troostite, and bainite) are recognizable by the metallographer.

When micromorphological studies fail, the investigator then proceeds to the electron microscopy for topography, to the EBP, ESCA, or SEM with energy dispersive X-ray analysis (EDXRA) for elemental analysis.

7.5.1. Topography

As indicated, topography of a surface greatly affects wear, friction, reflectivity, catalysis, and a host of other properties. Many techniques can be used to study surfaces, but most begin with visual examination supplemented by increasing magnification of the light microscope. Straightforward microscopy may be supplemented by either sample preparation techniques or specialized microscope accessories.

7.5.1.1. Dark-Field Illumination

There are two general methods of observing surfaces: dark field and bright field. Each of these, however, can be obtained with transmitted light from a substage condenser and with reflected light from above the preparation. For bright-field top lighting, the microscope objective itself must act as a condenser for the illuminating beam. For dark-field transmitted light, the condenser numerical aperture (NA) must exceed the NA of the objective, and a central cone of the condenser-illuminating beam, equal in angle to the maximum objective angular aperture, must be opaqued.

The stereobinocular microscope (stereo) is an arrangement of two separate compound microscopes, one for each eye, viewing the same area of an object. Because each eye views the object from a different angle, separated by about 14°, a stereoimage is obtained. The physical difficulty of orienting two high-power objectives close enough together for both to observe the same object limits the NA to about 0.15 and the magnification to about 200×.

The erect image is, however, an advantage, and solution to most surface problems starts with the stereo. There is ample working distance between the objective and preparation, and illumination is flexible. Many stereos permit transmitted illumination, and some permit bright-field top lighting. At worst, it is possible to shine a light down one body tube and observe the bright-light image with the second body tube. This is better accomplished with a trinocular head that uses a beam splitter rather than beam switcher.

The resolution of a stereo is only 2 μm, 20 times larger than the limit of a monoobjective microscope. Unfortunately, increased resolution results in a smaller working distance and a smaller depth of field, as shown in Table 7.4. As a result, it becomes more difficult to reflect light from a surface using side spotlights as the objective NA increases. In any case, the angle between the light rays and the surface must decrease rapidly as the NA increases and the working distance decreases. The surface should be uncovered, *i.e.*, no cover slip, and consequently all objectives with NA > 0.25 should be corrected for uncovered preparations (metallographic objectives).

Side *spotlights* work very well for 10-mm 10 × objectives and surprisingly well for higher power objectives, including 4-mm 45× objectives. Apparently enough light is reflected back and forth between the bottom lens of the objective and the preparation, so that a portion reaches the area in the field of view. This latter kind of illumination has to be recognized when using high-power transmitted light to avoid misinterpreting of structures. Thin metal flakes often look transparent because of unsuspected light reflected from the top surface. This sometimes occurs with no intentional top light at all if a nearly full condenser NA allows light to hit the front objective surface and be reflected back into the preparation.

Spot illuminators must be capable of concentrating all of their light onto a

Table 7.4. Examples of Working Distance Resolution and Depth of Field as a Function of Objective Numerical Aperture

Microscope	NA	Resolution (μm)	Working Distance (mm)	Depth of Field (μm)
Stereobinocular	0.14	2.20	50–90	15
Monoobjective	0.65	0.47	0.7	1

Source: Kane *et al.* (1974).

small spot a few millimeters in diameter to reflect sufficient light from the specimen for visual examination. The amount of reflected light is usually measured through the microscope for photomicrography by placing a small piece of white bond paper in the plane of the preparation. An exposure sufficient to just saturate the paper properly renders the various colors and shades of gray in any preparation replacing the paper.

Another way of using spotlights when the lamp itself obtrudes is to move it to one side, then direct the light beam to a small mirror near the objective, small mirrors like those used by a dentist are excellent for this purpose. The principle here is simply that when a mirror is concave, it can help concentrate a light beam.

One of the most significant variables when using unilateral illumination is the angle between the illuminating beam and the surface. The problem of directing light onto the specimen by spotlight or reflector because the working distance is too small can be solved by using reflection objectives. Both 8-mm, $20\times$ and 4-mm, $40-45\times$ reflecting objectives are available with working distances as great or greater than that of the refracting 16-mm, $10\times$ objective, *i.e.*, 5–6 mm. Very adequate dark-field top light can be achieved with reflecting $20-45\times$ objectives using either spotlights or reflectors.

Another way of "flattening" a surface involves using several spot illuminators (or reflectors) or fully annular illumination. There are various ways of accomplishing this. One simple way involves using a ping-pong ball, as proposed by Albertson (1964). He cut a ping-pong ball in half, then cut a small circular hole in the center of one hemisphere. He placed the *light-tent* over the preparation equator-side down and centered the objective through the north pole opening. At least two spotlights were then focused on the outside of the hemisphere. A soft diffuse illumination thereby pervaded the specimen area, lighting highlights and generally flattening the object.

The *Lieberkuhn illuminator*, a simple circular concave mirror fitting around the objective, has also been used for annular top light. Introduced many years ago by Lieberkuhn, it is still a useful device. To use it, the substage condenser must be removed so that parallel light rising around the specimen is reflected by the mirror onto the specimen.

The *Silverman ring illuminator*, is still available from E. Leitz. It is a circular array of small light sources, again fitting around the objective so that light is directed downward from all sides onto the surface.

The *fluorescent ring illuminator* is another variation on this theme. It is a ring fashioned from fluorescent tubing of the size used in advertising signs, manufactured by Artisto Grid Lamp Products. Any neon sign maker can also produce one of these with an appropriate transformer and a green-fluorescing gas. The tubing should be a 1–1.5 in. circle.

E. Leitz produces vertical illuminators for dark-field illumination under the name *Ultropak*. They are available in all magnifications from $3.8-75\times$; the latter is an oil-immersion objective. All Ultropak objectives use an annular mirror above the objective surrounding the objective optical path and at a 45° angle to reflect light from an external illuminator downward around the objective to the specimen. The low-power Ultropaks have an annular lens around the objective to focus annular light rays onto the specimen plane. The higher power Ultropaks use an annular mirror instead of a lens to focus incident light.

7.5.1.2. Bright-Field Illumination

A convenient transition from dark-field to bright-field illumination is the Episystem. Used by several manufacturers, it is a vertical illuminator with a double-mirror system. In one configuration, the Epi-Condenser is like the Ultropak, a dark-field vertical illuminator. However, by sliding a half-silvered mirror into position in the objective light path, some light is reflected downward through the objective onto the specimen. Some light reflected from the specimen under either of these conditions (central or annular mirror) reenters the objective, and some of this proceeds through the imaging process to the eye.

The annular mirror is, of course, a dark-field system; scratches on a polished metal surface, for example, appear white on a dark field. The central mirror, on the other hand, is a bright-field system, and scratches on a polished metal appear dark on a bright field. Other vertical illuminators are capable only of bright-field illumination. These are simply objectives with a thin glass plate at 45° just above the top lens of the objective and with a side tube for an illuminating beam.

A crude but effective bright-field vertical illumination effect for at least low-power objectives can be achieved by supporting a small fragment of clean cover slip between the specimen and the objective. The cover slip must be small enough to fit entirely within the working distance of the objective and large enough to cover the entire light path. The cover slip fragment is held in a small flattened bead of modeling clay either on the specimen or better on the objective itself. A $10\times$ 16-mm objective is the highest power lens that can be used for this purpose without seriously deteriorating the image because the cover slip thickness when traversed at a 45° angle is far too thick (0.24 mm for a 0.17-mm cover slip) for good imaging. The 16-mm objective must of course be corrected to use with a cover slip. Metallographic objectives are corrected for zero cover-slip thickness.

Another simple way of obtaining bright-field vertical illumination involves delivering light to the specimen through one body tube of a stereo.

Another top lighting system uses the $2\times$ auxiliary magnifier furnished as an accessory for the Wild M-5 stereo. This lens has a concave front face (facing the specimen), therefore, it acts like a Lieberkuhn reflector, so that light transmitted from the substage mirror is reflected from the large-diameter front lens onto the specimen. This gives adequate light for bright-field surface study.

When surface detail is not readily visible because contrast is low, phase contrast is a useful means of enhancing contrast. Phase contrast enhances optical path differences, and since surface detail generally involves differences in optical path (differences in height), these differences are made more apparent to the eye.

Reflected-light phase contrast is not used nearly so much as it was before the introduction of reflected-light Nomarski interference contrast (NIC). The latter also enhances optical path differences but without the halos that mar the phase contrast images. Although limited in resolving power compared to the light microscope on which it is based, the NIC gives fully as much information (and as clean micrographs) as the SEM.

The metallograph is a one-purpose instrument designed to give the best possible visual and photomicrographic image of an opaque surface. It is nearly always an inverted microscope system allowing easy access to a large unencumbered stage. All optics are below this stage and corrected for uncovered preparations. Generally, either bright-field or dark-field illumination is possible. Rigid construction helps minimize vibration, and metallographers find the metallograph almost indispensable for the routine examination of polished metal surfaces, it can be used to examine surfaces of any materials. There seems to be a tendency to abandon the metallograph for the universal microscope, which has greater flexibility with the addition of such accessories as NIC.

It is advantageous to have a single microscope–photomicrographic camera system with the highest quality accessories for all types of surface study and for photomicrography at any stage. All major microscope manufacturers produce such equipment; the price approaches that of a low-priced SEM.

7.5.2. Sampling Treatment Procedures Used to Enhance Contrast

The surface of any transparent, colorless (or nearly colorless) multicomponent substance, *e.g.*, paper, particle-filled polymers, porcelains, and some ceramics, is difficult to study and virtually impossible to photograph by LM. So much light penetrates the surface and is refracted, then reflected to the observer that the surface itself is lost in glare. This problem can be very easily solved by evaporating a thin film of metal onto the surface. The metal (usually aluminum, chromium, or gold) may be evaporated under vacuum in straight lines at any angle to the surface, from grazing to normal incidence; an angle of about 30° is often used. Under these conditions, the heights of surface elevations can be calculated from shadow lengths. Normal incidence is usually used for photomicrography of, say, a paper surface or when the surface is to be studied by another viewing procedure, such as interferometry or Tolansky's surface-profile microscope (Tolansky, 1952).

Transparent film replicas of opaque surfaces can be studied by transmission LM. This leads to the possibility of using transmission phase contrast or interferometry and the best possible optics, *e.g.*, oil-immersion planapochromats to study surface detail. In addition to these obvious advantages, replication is almost the only way of studying hard-to-reach surfaces, *e.g.*, the inner ball race of a ball bearing. Finally, replication not only gives a faithful rendition of surface detail but often picks off individual particles or coatings from the surface that may be related to the problem under investigation, *e.g.*, scoring of the ball race. The position of the particles relative to the surface geometry is also preserved by replication.

The best replication procedure involves coating with collodion solution followed by stripping the dried replica. With this procedure, surface debris becomes imbedded in the collodion and is held securely in place on drying; it is removed by stripping off the film. The replica is flattened on a microscope slide, so that an entire curved surface can be viewed microscopically in one focal plane.

A surface profile can be examined directly by taking a cross section and placing the surface on edge for microscopical study. This usually involves mounting the piece in a cured polymeric resin mount, then grinding and polishing down to the desired section. If the surface is soft, *e.g.*, copper, silver, polymers, *etc.*, it can be protected during polishing by electroplating or evaporating a thin coat of nickel metal over the surface before mounting and sectioning.

An interesting variation of this sectioning procedure involves taking the cross section at an angle other than normal to the surface, which magnifies the heights of elevations; for example, an angle of 5.7° from the surface gives a $10\times$ magnification in the vertical direction. Moore (1948) terms this a taper section; it is not, of course, restricted to cross sections of metals.

7.5.3. Chemical Composition and Solid-State Structure

7.5.3.1. Morphological Analysis

Characterization of a surface includes not only topography but chemical composition and solid-state structure. An experienced microscopist can identify many microscopic objects by shape, size, surface detail, color, luster, *etc.*

Unfortunately, it is difficult to describe a surface so that it is recognizable by someone else. Useful descriptive terms for surfaces include angular, cemented, cracked, cratered, dimpled, drusy, laminar, orange peel, pitted, porous, reticulated, smooth, striated, and valleyed.

A microscopist should recognize a paper as rag, mechanical coniferous, chemical nonconiferous, or as combinations of these by studying the paper surface. He/she should recognize such surface features on small particles as scales on wool, crossover marks on silk, striations on viscose rayon, twin bands or calcite, melt and crystal patterns on micrometeorites, lamellar steps on mica, the fibrous structure of incinerated wood sawdust, *etc.*, which help identify that substance.

Measurements of reflectance on polished surfaces can be used to calculate refractive indices of transparent substances and to give specific reflectance data for opaque substances, methods are discussed in detail by Cameron (1969). Reflectance and microhardness data are tabulated by Bowie *et al.* (1958) in a system for mineral identification.

The most sophisticated approach to identifying substances morphologically involves using metallography. Polished and etched metal sections show details of structure a metallographer can use to identify a metal or alloy and the heat treatment or other metallurgical processing that metal has undergone. Similar procedures of polishing and etching can be applied to other substances, *e.g.*, explosives (McGrath, 1969). It is possible to tell quickly whether the explosive was cast, pressed, or extruded; whether it is a single component or mixture; and with training, its components. Preetching with dilute acetone helps bring out the structure.

7.5.3.2. Staining Surfaces

Staining a surface, either chemically or optically, helps differentiate parts of a composite surface and identify the various phases. A variety of stains are available for diverse surfaces. Mineral sections can be etched with hydrofluoric acid then stained with $Na_3Co(NO_2)_6$ to differentiate quartz (unetched), feldspars (etched but unstained), and potassium feldspars (etched and stained yellow). Paper surfaces can be stained with Harzberg stain (Calkin, 1934) to differentiate lignified cellulose (*e.g.*, straw, manilla, and mechanical wood pulp [yellow or yellow green]) from purified cellulose (*e.g.*, chemical wood pulp, bleached straw, or manilla [blue or blue violet]) or pure cellulose (*e.g.*, cotton or linen [wine red]). Isings (1961) selectively stains unsaturated elastomers with osmium tetroxide.

There are a variety of ways of building up thin films on surfaces to show interference colors when the right thickness is attained. Furthermore, the thickness may vary from grain to grain or composition to composition and provide a color guide for phase identification. One of the simplest procedures was developed by Staub *et al.* (1968). They applied a drop or two of 0.75% Parlodion in amyl acetate over the polished surface and allowed the specimen to drain and dry in a vertical position. This method creates different film thickness over the different phases, thus leading to differentiation from interference colors. The procedure works best for metals and opaque ores.

Any metal that can be anodized usually develops oxide coatings over each grain of a thickness depending on the composition and orientation of that grain. Transparent oxide films then show interference colors, which differentiate one grain from another. Picklesimer (1967) illustrates this technique for alloys of zirconium and niobium; he also discusses older ways of developing thin film colors on metals both electrolytically and by heating.

Allmand *et al.* (1970), among others, have vacuum-evaporated thin films of highly refractive transparent substances, *e.g.*, TiO_2 or ZnS, onto a polished, etched metal surface. As the film builds up to different thicknesses over different solid phases, each film is differentiated by interference colors.

To delineate magnetic domains McKeehan *et al.* (1934) placed a drop of a dilute suspension of submicrometer particles of magnetic iron oxide Fe_3O_4 on the surface. The oxide particles migrate to the surface, they form a pattern accurately mapping the domain structure.

These optical and chemical staining procedures are often necessary to develop sufficient contrast in a specimen so that automatic image analysis can take quantitative measurements of composition and grain size.

7.6. ELECTRON MICROSCOPY

There is hardly a field in materials science where the physical nature of the surface is not an important feature. For example, in fatigue fracture, cracks nucleate at the surfaces of materials, and the rate at which they nucleate is greatly influenced by the detailed topography of the surfaces. In the field of thin-film devices, the manufacturing tendency has been to reduce the size of electronic components. Surface-to-volume ratios are now exceedingly high, and as Young (1971) points out, we are not far from the point where we can anticipate devices employing single layers of atoms. However, the device industry, which presently employs films in the 10–100 Å range, suffers very high failure rates due to surface imperfections, stacking-fault intersections, voids in films, thermally induced pits, and multiple steps. As a result of these deficiencies, considerable resources have been employed to control imperfections by closely controlling processing variables. In other areas, elaborate polishing, cleaning, and smoothing techniques have been developed in an effort to eliminate the variability associated with surfaces. However, none of these efforts can improve on a detailed knowledge of the actual surface topography.

7.6.1. Transmission Electron Microscopy

Electron microscopy is useful for studying pigments, particle size, distribution, and surfaces where very high resolution is required. Table 7.5 shows the most used electron microscopy methods together with capability, magnification, and resolution. Our discussion here focuses on the transmission electron microscope, which, like the ordinary optical microscope, simultaneously illuminates the whole specimen area and employs Gaussian optics to generate the image. For a comparative review of the capability of other kinds of topographic measurers, see Young (1971); for a discussion of the flying spot and other types of instruments, see Johari (1974). Unlike the transmission electron microscope, the scanning microscope, its most serious competitor, at least in terms of number, illuminates only one spot on the specimen at a time and forms its image sequentially. The transmission electron microscope has greater resolving power than an equivalent scanning microscope (as is generally true of types that employ Gaussian optics), and it spreads the illumination over the whole specimen rather than concentrating it in one high-density spot. Consequently, the scanning microscope must employ a much smaller beam current than the transmission electron microscope and in my experience, causes much less overall specimen damage than the transmission electron microscope in highly susceptible materials, such as polymers. On the other hand, when working with metals and regular accelerating voltages (100–150 kV) and equipped with a good decontamination device, the transmission electron

Table 7.5. Electron Microscopy Instruments and Capabilities

Instrument	Capability	Magnification (X)	Resolution (A)
Transmission electron microscope[a]	Internal structure	~1.5MX[b]	~3[b]
Scanning electron microscope	Topological structure	40X–300KX	15
Auger electron microscope	Topology structure	~5K[b]	250

Notes: [a]TSEM is a mode of the transmission electron microscrope.
   [b]Approximate value, see references in text.

microscope, can operate virtually ad infinitum without seriously deteriorating the area under observation. The same is hardly likely in the case of the scanning microscope unless it is also equipped with a good decontamination device.

Flying-spot instruments permit point-by-point analysis of surface properties. At first, it would appear that transmission electron microscopes, which illuminate an entire sample, would not be suitable for such an application, and in general, this is so. However, a new transmission electron microscope named EMMA 4 has been developed with combined transmission electron microscope and probe capability by introducing a minilens into the illumination system (Cook *et al.*, 1969; Jacobs, 1971). The EMMA 4 has demonstrated considerable power in a number of applications and could easily be applied to surfaces, but it will not be further considered here because our primary emphasis is on the topography of paint.

A great advantage of the scanning microscope is its ability to deal with bulk specimens. Unfortunately, nonconducting samples have to be given a light coating of metal, typically gold; otherwise, charging effects seriously impair the resolution of the image. Transmission electron microscopes are not subject to this limitation, and the techniques described here apply universally to all materials. Such a statement is of course theoretical because numerous practical problems beset the preparation of all kinds of materials for observation in the transmission electron microscope.

In the transmission electron microscope, electrons forming the image must pass through the specimen; thus, the specimen thickness is limited to a few thousand angstroms, or microns for the fortunate owners of a high-voltage instrument. There are two approaches to studying the surfaces of solids: In one approach, a replica of the surface can be made—for example, a carbon replica made by vacuum depositing a 100–1000-Å film on the surface—and carefully removed by some etching technique, then mounted in the microscope. While the image obtained from such a replica represents the surface topography, it is frequently subject to distortion and artifacts and often difficult to interpret. Moreover, the process of replication seriously reduces the resolution ultimately obtainable with the microscope. In the other approach, it is necessary to plate a suitable material onto the surface of interest and then to section a thin slice normal to that surface. The section is then mounted for observation in the microscope and the surface is viewable in profile. The resolving power of the instrument can be fully exploited by the profile method, and it has the additional advantage of revealing the surface topography in relation to the underlying structure of the material. Alternatively, to obtain a bird's-eye view of a surface rather than the one-dimensional profile of a section, it is feasible for many materials to protect the surface of interest with a nonconducting layer and then to electropolish the sample from the other side. Local thinning near the surface enables an observer to see rather large areas of the surface if the nonconducting layer can be dissolved or washed away in a suitable solvent.

This subject is too broad to permit detailed description of any kind of microscope or of the theory by which it is employed. Since many excellent books have been written on the microscope itself (Klemperer, 1953; Thomas, 1962; Haine *et al.*, 1961; Heidenreich, 1964; Grivet, 1965; Hirsh *et al.* 1965; Amelinckx, 1964, 1970; Hall, 1966; Wyckoff, 1949) on methods of preparing specimens (Wyckoff, 1949; Kay, 1961; Thomas, 1971), and on the theory of contrast (Heidenreich, 1964; Hirsh *et al.*, 1965; Amelinckx, 1964, 1970), only a very brief description is provided of contrast principles, specimen preparation methods and applications where replication and sectioning techniques have been successfully employed to study surfaces.

7.6.1.1. Contrast Theory

The problem now is to interpret the electron images obtained by the two approaches available for studying surfaces: the replication and profile methods. Since electrons pass through the samples, the images formed from them are going to be strongly affected by the interaction of the electrons with the sample material. The atomic spacings of most materials and the wavelengths of the electrons obtained from the accelerating voltages employed are conducive to diffraction. Both elastic and inelastic scattering is possible with the inelastic type dominating in thick samples (>2000 Å for 100-kV electrons) and the elastic type more important for readily usable thicknesses. Many different types of inelastic scattering occur (Hirsh *et al.*, 1965; Amelinckx, 1964, 1970): plasma losses, photon interactions, bremsstrahlung radiation, *etc.* The net effect is that some of the incident electrons are deflected from the collimated, axially parallel beam focused on the specimen by the illumination system. These deflected beams are focused at different points in the back focal plane of the objective lens. To obtain contrast in the image, an objective aperture is inserted in the back focal plane to block the scattered beams and permit only the direct beam to form an image in the projection lens system of the microscope. This image is called the bright-field image, and its details are determined by the extent to which scattering has occurred in different regions of the specimen. Alternatively, a dark-field image can be formed by shifting the objective aperture laterally to block the direct beam and permit only one of the scattered beams to pass into the image system of the microscope. The bright- and dark-field images reveal many details about imperfections contained in the specimen or on its surface.

Although this method of obtaining contrast is quite general, the scattering processes involved vary widely for different materials; thus, it is convenient to differentiate those that occur in the two approaches used for studying surfaces. In the replication method, most replicas are essentially amorphous. The diffraction of electrons from replicas is therefore different from the type that occurs in profile sections, which are more likely to be crystalline. In replicas, the diffraction patterns (*i.e.*, the distribution of electron intensity in the back focal plane) are rather hazy, with a fairly high intensity scattered at a Bragg angle corresponding to the most populous interatomic spacing. Since the structure is generally uniform, intensity distributions in the electron images are also uniform unless the thickness of the replica varies. The scattering amplitude must increase where the thickness of the replica is greater, due to a greater probability of scattering events. Therefore, such regions appear darker in relation to the background in a bright-field image. Clearly, if a uniformly thick replica is taken from a surface, a step will produce a region two or more times thicker than average, and inclined regions will have their effective thickness increased by the secant of the angle of inclination. Heidenreich has worked out in detail the contrast to be expected from such specimens (Heidenreich, 1964).

Materials used for replication, such as carbon, are usually so transparent to electrons that small thickness variations produce no observable contrast. It is customary, therefore, to enhance contrast by shadowing the replica with a heavy metal, which produces marked variations in contrast. In addition, shadows help accentuate height differences in the specimen and open the way to obtaining quantitative information about the surface topography via stereo-micrometry.

For profile specimens, the ordered nature of the crystals gives rise to marked elastic scattering of the incident beam. If the specimen is mono-crystalline, the diffraction pattern is a spot pattern, readily identifiable by techniques described in much more detail elsewhere (Hirsh *et al.*, 1965). Since the theory of electron diffraction is well understood, detailed quantitative information can be obtained from the specimen by tilting it in seriatim to different orientations and exciting a variety of Bragg reflections (Heidenreich, 1964; Grivet, 1965). This information can be obtained about both the crystal-lography of the specimen and its internal defects. Although both the specimen to be observed in profile and the material plated on the surface to preserve its outline can be uniform in thickness, the contrast is entirely different from that observed in a replica. Small deflections inevitably occur in the ordered atomic arrangements of the crystals, either by elastic strains (either gross or localized to defects) or by variations in composition. Since electron wavelengths are generally quite small in relation to atomic spacings, Bragg's law indicates that the scattering angle of the diffracted beams should also be small, $\sim 1°$. Consequently, even small elastic strains in a specimen can locally distort the crystal away from the Bragg condition and cause large variations in scattering intensity and therefore in image contrast.

Interpreting the Gaussian image requires detailed knowledge of diffraction theory. The kinematical theory of diffraction, familiar from X-ray theory, is not suitable for electron microscopy because this theory requires only a small proportion of the incident electrons to be scattered. The most interesting effects, however, are observed at crystal orientations near Bragg reflections, where large intensities are scattered and the kinematical theory no longer applies. Instead, observers turn to the dynamic theory of diffraction, formulated either in terms of wave mechanics or wave optics. The wave mechanical formulation is much more convenient, and the mathematics employed are relatively simple (Hirsh *et al.*, 1965; Amelinckx, 1964, 1970). In this theory, allowance is made for large scattering intensities. Further, if the crystal orientation satisfies the Bragg condition for the direct beam, then it also satisfies it for the diffracted beam. Consequently, the diffracted beam is doubly diffracted back into the direct beam. If the crystal is thick enough, and it usually is, multiple reflections occur. Dynamic theory takes such reflections into account, as well as two other important factors. Electrons interact with the potential of the crystal, which varies periodically with the atomic array. Electrons between atom rows experience a higher potential energy than electrons in the neighborhood of the atoms. Since electrons are given a fixed amount of energy by the electron gun, those electrons with higher potential energy have a lower kinetic energy and therefore a different wavelength. Consequently, electron beams within the crystal vibrate together. This interference is normally observed as fringe patterns at inclined interfaces within the specimen (grain boundaries, twins, stacking faults, precipitates, and any other planar defects). It permits observers to see such interfaces if they tilt the specimen close to the Bragg condition where the interference effects are optimized. Finally, the dynamic theory takes into account absorption losses of electrons, which constitute another source of informative imaging phenomena.

7.6.1.2. Techniques

Replication techniques have been developed to a considerable degree of sophistication, comprising both one-stage and two-stage methods. These techniques use a wide variety of replicating materials, depending on the application (Kay, 1961). Plastic replicas have a serious resolution limitation, because the molecule of the plastic itself may be larger than the resolving power of the instrument. This means that the aggregate of the replica can interfere with the fine details of the surface of interest. Consequently, shadowed carbon replicas, which have much better resolution, are used almost exclusively in the most exacting work. They are generally prepared by a two-stage method: The surface of interest is coated with a Formvar (or equivalent) film by flooding with a 2% solution of the plastic in chloroform and draining away the excess. When the film is dry, it is scored into squares, backed with Scotch tape, and stripped for evaporation of carbon onto the surface structure of the film. Then, 10–20 nm of carbon are deposited along with a lighter deposit of shadowing metal, which can be applied before, during, or after evaporation of the carbon. Finally, the plastic film is dissolved by washing it in a solvent, such as acetone, and carbon replicas are netted out for mounting in the microscope. Although the necessary procedures and techniques are supplied in the standard texts (Hall, 1966; Wyckoff, 1949; Kay, 1961), considerable practice is required to obtain good results, and details vary in relation to the practitioner's skill. The most important factor seems to be the thickness of the plastic film. The smoother the surface to be replicated, the thinner the plastic film should be, with the main objective to prevent disintegration of the carbon replica when the plastic is washed. Before the plastic dissolves, it expands, and strains transmitted to the carbon replica can be destructive. For very rough surfaces, such as fractures, the corrugations in the carbon replica strengthen them, and thicker plastic coatings can be used.

While a carefully prepared replica should accurately reflect horizontal distances in the surface of interest, greater doubts may exist about how well vertical dimensions are represented. Growth features in the vertical direction collapse and rumple, but wide experience with powder samples, surfaces with slip steps, and fracture surfaces, has demonstrated that over short distances, vertical features of a surface are accurately replicated and retained in the microscope. Thus, stereomicroscopy, which permits the surface to be viewed and measured directly in three dimensions, is regarded as a valid technique.

The value of routine stereoscopic observation as an aid in understanding complicated surfaces does not appear to be widely appreciated, however, the depth definition afforded by the technique gives a clearer overall view and shows up otherwise hidden details, just as aerial stereophotographs are used to spot camouflage. Thomas (1971) and others (Howie, 1965; Nankivell, 1963), have pointed out how easy it is to obtain stereo pictures. Two different views of the same specimen area are required at different tilts. Depending on the amount of stereo effect desired at the final viewing magnification, the relative tilting angle is 10–20°. By tilting the replica at equal angles around its horizontal position, the stereo view appears to lie horizontally. For depth measurements, the angle of tilt between the two views Φ and the magnification M should be accurately known. The height differences Δh between vertically spaced points in the image can be measured from the parallax Δp between the corresponding pairs of image points in the two views of the stereo pair, using the relations $\Delta h = \Delta p[(2M \sin[(\Phi/2)]])$ (Thomas, 1971). The parallax is measured by a stereomicrometer placed on the photographs. Complete details about how to use such an instrument and how to position and orient the micrographs with respect to the stereoviewer can be obtained from the various manufacturers' handbooks (*Handbook for Wild ST4 Mirror Stereoscope*, 1967). Complete surface profiles can be obtained by such techniques and contour maps drawn; however, a very large number of measurements is required to obtain contours, and the procedure is exceedingly tedious. In spite of the reluctance observers may accordingly feel, a few examples can be found in the literature (Boyde, 1970; Bradley *et al.*, 1956; Halliday, 1961; Jones *et al.*, 1970). Useful hints and general experience may be gained from stereo applications in scanning microscopy and reflection electron microscopy (Boyde, 1970; Bradley *et al.*, 1956; Halliday, 1961; Jones *et al.*, 1970).

7.6.1.3. Transmission Scanning Electron Microscopy (TSEM)

Although most commercial SEM are used to study surface features, signals transmitted through thin samples can be collected by a suitable detector placed beneath the sample; thus, SEM can be used in the transmission mode (TSEM). Comparing the TSEM with a conventional transmission electron microscope reveals that the two microscopes are equivalent, so that information obtained from a transmission electron microscope can theoretically also be obtained from a TSEM (Jones *et al.*, 1970; Zeitler, 1971). Dark-field and diffraction work are possible for metallurgical and materials science applications, although in commercial SEMs the resolution (approximately the beam diameter—about 50–75 Å, depending on the thickness, primary kV, and other parameters) is much poorer than that of transmission electron microscope about 2–5 Å. The main advantage is in biological work: (1) Thicker samples can be examined in the STEM than in the transmission electron microscope, because scattered electrons are collected point by point and amplified, unlike the transmission electron microscope, where primarily coherent electrons are collected and the image is magnified through an electron optical system; in other words, TSEM is similar to a built-in image intensifier. (2) Examination at low magnifications is possible, and since surface examination is also possible, correlation with other results, such as those from various types of optic microscopy, is easily possible. (3) Since the SEM can run at a much lower voltage (5–25 kV, typically), specimen damage is reduced compared to a transmission electron microscope (typically run at 50–100 kV). (4) Since specimens are not physically placed in the field of the final lens, much larger specimens can be used for the TSEM than for the transmission electron microscope.

Specially built TSEMs with a field-emission source and an ion-pumped vacuum system have been used by Crewe to obtain point resolutions of 5 Å and to resolve atoms of uranium (Crewe, 1970).

7.6.2. SEM

A detailed examination of material is vital to any investigation related to processing properties and the behavior of materials. Characterization includes information about topographical features, morphology, distribution, identification of differences based on chemistry, crystal structure, physical properties, and subsurface features.

Before the advent of the SEM (Johari, 1971), several tools, such as the optic microscope, the transmission electron microscope, the electron microprobe analyzer, and X-ray fluorescence, were employed to accomplish partial characterization; this information was then combined for a fuller description of materials. Each of these tools has proficiency in one particular aspect and complements the information obtained with other instruments. The information is partial because of the inherent limitations of each method, such as the invariably cumbersome specimen preparation, specialized observation techniques and interpretation of results.

The SEM serves as a bridge between the optic microscope and the transmission electron microscope, although the TSEM approaches the resolution and magnification obtainable by the transmission electron microscope. The SEM has a magnification of 3–100,000×, a resolution of about 200–250 Å, and a depth of field at least 300 times or more the optic microscope, which results in the three-dimensional high-quality photographs of coating and pigments. Because of the large depth of focus and large working distance, the SEM permits direct examination of rough conductive samples at all magnifications without special preparation. All surfaces have to be coated with a thin conductive layer of carbon, gold, or palladium, *etc.* All electron microscopy instruments are strictly topological viewing tools (*i.e.*, only the immediate surface is visible).

7.6.3. X-Ray Spectroscopy in the SEM (EDXRA)

The SEM has so many material characterization capabilities that it is often considered the ideal tool for material characterization (Johari, 1971; Howell *et al.*, 1972; Boyde, 1970), however, X-ray spectroscopy (Johari *et al.*, 1974) tremendously enhances the analytic value of the SEM in material characterization by providing chemical analysis of the sample along with surface topology.

In the wavelength diffractometer (WD) method, a crystal of a known spacing d separates X-rays according to Bragg's law, $n\lambda = 2d\sin\theta$, so that at a diffraction angle θ X-rays of specific wavelengths are detected. To cover the whole range diffractometers are usually equipped with many crystals even then, considerable time is needed to obtain an overall spectrum of all elements present. The resolution of the crystal in separating X-rays of different wavelengths is very good (of the order of 10 eV), but efficiency is very poor. To improve collection efficiency, curve crystal fully focusing diffractometers are used.

For nondispersive (ED) spectrometers, the energy of an incoming X-ray photon is converted into an electric pulse in a lithium-drifted silicon crystal. A bias voltage applied to the crystal collects this charge, which is proportional to the energy of the X-ray. This pulse is amplified, converted into a voltage pulse, and fed into a multichannel analyzer. The analyzer sorts out the pulses according to their energy and stores them in the memory of the correct channel. The resulting spectrum can be displayed on a cathode-ray tube (CRT), plotted on a chart, or printed out numerically.

Characteristic X-rays emitted under the effect of the electron beam provide information about the nature and amount of elements present in the volume excited by the primary beam. Energy-dispersive X-ray analysis attachments, consisting of a lithium-drifted silicon crystal, a multichannel analyzer, and necessary electronics, are finding increasing use on many SEM models. This method is capable of detecting elements with atomic number as low as 9 (fluorine) in an SEM and 5 (boron) in the TSEM, with a detectability limit of 0.5% by volume. A spectrograph of elements is generated, and it can be presented on a CRT, printed graphically for a permanent record, or stored on magnetic disk. In a spectrograph, the X-y plot consists of wavelength versus intensity, and the area under the peaks is indicative of the amount present. Wavelength diffractometers, used with electron beam probe microanalyzers, are also available as an accessory on the SEM. Depending on the need and amount of funds available, one or both of these systems up to any level of sophistication can be provided.

7.6.4. Auger Electron Microscopy

This technique is most powerful and analyzes the first few atom layers (10 Å or less) on the surface of the sample. Many applications include detecting very low concentrations of impurity segregation, which critically affects mechanical properties of some body-centered cubic metals and alloys.

Auger electron analysis (AES) with the SEM requires appropriate energy-analyzing equipment but more importantly, a much better vacuum system than is currently available in most instruments. Since AES analyzes only the first few surface layers, samples must be free from any surface film. The necessary energy analysis equipment is available as standard commercial items. Standard Auger spectra are available for all elements for various Auger transitions. Overlapping spectra from two elements may create some problems of elemental separation, but with high-resolution energy analysis equipment, procedures similar to those used with X-rays can successfully be employed.

Specimens examined in the SEM mode must be coated with a conductive layer similar to the process in conventional SEM instruments to avoid specimen charging; see MacDonald (1971) and Chang (1971) for excellent review articles on AES.

8

Methods of Analyzing Vehicles

8.1. INFRARED SPECTROSCOPY

Infrared spectroscopy is an excellent tool for identifying components of vehicles. The infrared region of the electromagnetic spectrum extends from the red end of the visible spectrum to microwaves; that is, the region includes radiation at wavelengths from 0.7–500 μm, or in wave numbers, from 14,000–20 cm^{-1}. The spectral range of greatest use is the midinfrared region, which covers the frequency range from 200–4000 cm^{-1} (50–2.5 μm). Infrared spectroscopy (IR) involves twisting, bending, rotating, and vibrational motions of atoms in a molecule. On interacting with infrared radiation, portions of the incident radiation are absorbed at particular wavelengths. The multiplicity of vibrations occurring simultaneously produces a highly complex absorption spectrum, which is uniquely characteristic of functional groups constituting the molecule and of the overall configuration of the atoms as well. Suggested review articles on the fundamentals of infrared spectroscopy are Bellamy (1958), Colthup (1964), Gianturco (1965), Herzberg (1945), and Nakanishi (1962).

8.1.1. Molecular Vibrations

Atoms or atomic groups in molecules are in continuous motion with respect to one another. The possible vibrational modes in a polyatomic molecule can be visualized from an imaginary mechanical model of the system. Atomic masses are represented by balls whose weight is proportional to the corresponding atomic weight, arranged in accordance with the actual space geometry of the molecule. Mechanical springs, with forces proportional to the bonding forces of chemical links, connect and keep the balls balanced. If the model suspended in space is struck by a blow, the balls appear to undergo random chaotic motions. However, if the vibrating model is observed with a stroboscopic light of variable frequency, certain light frequencies are found at which the balls appear stationary. These frequencies represent the specific vibrational frequencies for these motions.

For infrared absorption to occur, two major conditions must be fulfilled. First, radiation energy must coincide with the energy difference between the excited and ground states of the molecule. Radiant energy is then absorbed by the molecule, increasing its natural vibration. Second, the vibration must entail a change in the electrical dipole moment, a restriction that distinguishes infrared from Raman spectroscopy.

Stretching vibrations involve changes in the frequency of the vibration of bonded atoms along the bond axis. In a symmetric group, such as methylene, there are identical vibrational frequencies, *i.e.*, the asymmetric vibration. In space, these two vibrations are indistinguishable and referred to as one doubly degenerate vibration. In the symmetric-stretching mode, there is no change in the dipole moment as the two hydrogen atoms move equal distances in opposite directions from the carbon atom, and vibration will be infrared inactive. There is a change in the dipole moment, since during these vibrations, the centers of highest positive charge (hydrogen) and negative charge (carbon) move so that the electrical center of the group is displaced from the carbon atom. These vibrations are observed in the infrared spectrum of the methylene group.

When a three-atom system is part of a larger molecule, it is possible to have bending or deformation vibrations, which imply movement of atoms away from the bonding axis; four vibration types are identified.

- Deformation or scissoring: Two atoms connected to a central atom move toward and away from each other, deforming the valence angle.
- Rocking or in-plane bending: The structural unit swings back and forth in the symmetry plane of the molecule.
- Wagging or out-of-plane bending: The structural unit swings back and forth in a plane perpendicular to the molecule's symmetry plane.
- Twisting: The structural unit rotates back and forth around the bond that joins it to the rest of the molecule.

In larger groups joined by a central atom, bending vibrations are split due to in-plane and out-of-plane vibrations; an example is the doublet produced by the *gem*-dimethyl group. Bending motions produce absorption at lower frequencies than fundamental stretching modes.

Molecules composed of several atoms vibrate not only according to the frequencies of the bonds but also at overtones of these frequencies; that is, when one tone vibrates, the rest of the molecule is involved. The harmonic (overtone) vibration possess a frequency that represents approximately integral multiples of the fundamental frequency. A combination band is the sum, or the difference between, the frequencies of two or more fundamental or harmonic vibrations. The uniqueness of an infrared spectrum arises largely from these bands, which are characteristic of the whole molecule. The intensities of overtone and combination bands are usually about one-hundredth of those of fundamental bands.

The intensity of an infrared absorption band is proportional to the square of the rate of change of dipole moment with respect to the displacement of the atoms. In some cases, the magnitude of the change in dipole moment may be quite small, producing only weak absorption bands, as in the relatively nonpolar C≡N group. By contrast, the large permanent dipole moment of the C=O group causes strong absorption bands, which is often the most distinctive feature of an infrared spectrum. If no dipole moment is created, as in the C=C bond (when located symmetrically in the molecule) undergoing stretching vibration, then no radiation is absorbed and the vibrational mode is said to be infrared inactive. Fortunately, an infrared inactive mode usually gives a strong Raman signal.

As defined by quantum laws, vibrations are not random events but can occur only at specific frequencies governed by the atomic masses and strengths of the chemical bonds. Mathematically, this is expressed as

$$\bar{\nu} = \frac{1}{2\pi c}\sqrt{\frac{k}{\mu}}$$

where ν is the frequency of the vibration, c is the velocity of light, k is the force constant, and μ is the reduced mass of the atoms involved. The smaller the mass of the vibrating nuclei and the greater the force restoring the nuclei to the equilibrium position, the greater the frequency. Motions of hydrogen atoms are found at much higher frequencies than motions of heavier atoms. For multiple-bond linkage, the first constants of double and triple bonds are roughly two and three times those of the single bonds, and the absorption position becomes approximately two and three times higher in frequency; interaction with neighbors may alter these values, as will resonating structures, hydrogen bonds, and ring strain.

Example: Calculate the fundamental frequency expected in the infrared absorption spectrum for the C—O stretching frequency. The value of the force constant is 5×10^5 dyn cm^{-1}.

$$\bar{v}(\text{cm}^{-1}) =$$

$$\frac{1}{(2)(3.14)(3 \times 10^{10})} \sqrt{\frac{(5 \times 10^5)(12 + 16)(6.023 \times 10^{23})}{(12)(16)}} = 1110 \text{ cm}^{-1}$$

8.1.2. Instrumentation

It is convenient to divide the infrared region into three segments, with the dividing points based on instrumental capabilities, because different radiation sources, optic systems, and detectors are needed for the different regions. The standard infrared spectrophotometer covers a range from 4000–650 cm^{-1} (2.5–15.4 μm). Although many prism instruments are still used, there has been an almost complete transition to filter-grating and prism-grating spectrophotometers. Grating instruments offer higher resolution, which permits separation of closely spaced absorption bands, more accurate measurements of band positions and intensities, and higher scanning speeds for a given resolution and noise level. Modern spectrophotometers generally have attachments that permit speed suppression; scale expansion; repetitive scanning; and automatic control of slit, period, and gain. Such accessories as beam condensers, reflectance units, polarizers, and micro cells can usually be added to extend versatility or accuracy.

Spectrophotometers for the infrared region are composed of the same basic components as instruments in the visible ultraviolet region, although sources, detectors, and materials used in fabricating the optical components are different except in the near infrared. Radiation from a source emitting in the infrared region is electronically regulated at a low frequency, often 10–26 times per second. It is passed first through the sample and then through the reference before entering the monochromator. This minimizes the effect of stray radiation, a serious problem in most of the infrared region. Temperature and relative humidity in the room housing the instrument must be controlled.

8.1.3. Radiation Sources

In the region beyond 5000 cm^{-1} (2 μm), blackbody sources without envelopes are commonly used. The same spectral characteristics cited for the tungsten incandescent lamp apply to these as well. Unfortunately, the emission maximum lies in the near infrared. A fraction of the shorter wavelength radiation is present as stray light, which is particularly serious for long-wavelength measurements.

A closely wound Nichrome helix can be raised to incandescence by resistive heating. A black oxide forms on the wire, which gives acceptable emissivity. Temperatures up to 2012°F (1100°C) can be reached. This source is recommended where reliability is essential, such as in nondispersive process analyzers and inexpensive spectrophotometers. Although simple and rugged, this source is less intense than other infrared sources.

A hotter, and therefore brighter, source is the Nernst glower, which has an operating temperature as high as 1500°C. Nernst glowers are constructed from yttrium-stabilized zirconium oxide in the form of hollow rods 2 mm in diameter and 30 mm in length. The ends are cemented to short ceramic tubes to facilitate mounting; short platinum leads provide power connections. Nernst glowers are fragile. They have a negative coefficient of resistance and must be preheated to be conductive. Therefore, auxiliary heaters must be provided as well as a ballast system to prevent overheating. A Nernst glower must be protected from drafts, but at the same time, adequate ventilation is needed to remove surplus heat and evaporated oxides and binder. Radiation intensity is approximately twice that from Nichrome and Globar sources except in the near infrared.

The Globar, a silicon carbide rod 4 mm in diameter and 50 mm in length, possesses characteristics intermediate between heated wire coils and the Nernst glower. It is self-starting, with an operating temperature near 2372°F (1300°C). The temperature coefficient of resistance is positive, and it can conveniently be controlled with a variable transformer. The Globar's resistance increases with the length of time used, so provision must be made to increase the voltage across the unit; the electrodes must be water cooled.

In the very far infrared beyond 200 cm^{-1} (50 μm), blackbody-type sources lose effectiveness, since their radiation decreases with the fourth power of wavelength. High-pressure mercury arcs, with an extra quartz jacket to reduce thermal loss, give intense radiation in this region. Output is similar to that from blackbody sources, but additional radiation is emitted from a plasma, which enhances the long-wavelength output.

8.1.3.1. Detectors

At the short-wavelength end, below about 1.2 μm, preferred detection methods are the same as those used for visible and ultraviolet radiation. Detectors used at longer wavelengths can generally be classified into two broad groups: (1) quantum detectors, which depend on internal photoconductive effects resulting from the transition of an electron from one valence band to a conduction band within the semiconductor receptor; and (2) thermal detectors, in which radiation produces a heating effect that alters some physical property of the detector. Quantum detectors are faster and more sensitive but severely restricted with respect to the range of wavelengths to which they can respond. Thermal detectors are usable over a wide range of wavelengths, actually over the entire spectral region in which the light-absorbing part of the detector can be regarded as black, but they suffer from relatively low sensitivity and slow response. The response time sets an upper limit to the frequency at which radiation can usefully be chopped, since the total mass represented by receiver, absorbing material, and temperature-sensing element must heat or cool during each half-cycle of chopping frequency.

The basic forms of thermal radiation detectors are the radiation thermo-couple, the Golay detector, and the bolometer. The thermocouple, the most widely used of all infrared detectors, is usually fabricated with a small piece of blackened gold foil (to absorb radiation) welded to the tips of two wire leads made of dissimilar metals chosen to give a large thermoelectric electromotive force (emf). Lead materials can be semiconductors, one with a large positive thermoelectric power with respect to gold and the other a large negative power. The entire assembly is mounted in an evacuated enclosure with an infrared-transmitting window so that conductive heat losses are minimized. Typical thermocouple detectors have a sensitive area of 0.5 mm^2, a response time of 40 msec, a dc resistance between $10–200$ Ω, a signal voltage of $0.1–0.2$ μV, and noise-equivalent power of 10^{-10} W at a chopping frequency of 5 Hz. To prevent faint signals from being lost in the stray signals (noise) picked up by the lead wires, a preamplifier is located as close to the detector as possible. The "cold" junction of the thermocouple actually consists of heavy copper lugs in contact with thermocouple sites. Since the detector needs to respond only to chopped radiation to give an ac output, only temperature changes are significant; hence, the actual temperature of the cold junction is unimportant. Receiver size is chosen to match a reduced image of the spectrometer's exit slit.

The Golay detector is pneumatic in principle. The unit consists of a small metal cylinder closed by a rigid blackened metal plate (2-mm square) at one end and by a flexible silvered diaphragm at the other end. The chamber is filled with xenon. Radiation passing through a small infrared-transmitting window is absorbed by the blackened plate. Heat, conducted to the gas, causes it to expand and deform the flexible diaphragm. To amplify distortions of the mirrored surface of the flexible diaphragm, light from a lamp inside the detector housing is focused on the mirror, which reflects the light beam onto a vacuum phototube. Motion of the diaphragm moves the light beam across the phototube surface and changes its output. In an alternative arrangement, the rigid diaphragm is used as one plate of a dynamic condenser; a perforated diaphragm a slight distance away serves as the second plate. Distortion of the solid diaphragm relative to the fixed plate alters the plate separation and hence the capacity. With either arrangement, the alternating output corresponds to the chopping frequency. Response time is on the order of 20 msec, corresponding to a chopping frequency of 15 Hz. The Golay detector has a sensitivity similar to that of a thermocouple. The angular aperture is 60°, so the unit must be used with a system of condensing mirrors to concentrate the radiation. The detector is significantly superior for the far infrared beyond 50 μm.

A bolometer is a miniature resistance thermometer usually constructed of metal or a semiconductor. The resistance of a metal increases with temperature about 0.19% per degree Fahrenheit (0.35% per degree Celsius), whereas that of a semiconductor decreases about 3.9% per degree Fahrenheit (7% per degree Celsius). A small flake of lightly doped germanium or silicon, cooled with liquid helium, constitutes a very effective bolometer. Two sensing elements as identical as possible, mounted close to one another, with one shielded from the radiant energy, form two arms of a Wheatstone bridge. For a thin slice of germanium cooled to 1.2 °K, a signal of about 10^{-12} W can be detected, which is two orders of magnitude smaller than what can be detected with a Golay cell. Another detector in the far infrared is made with a very pure piece of InSb. It is generally considered to be an electronic bolometer, where the electron gas can be heated for a short time without coupling to the lattice. Its sensitivity is comparable to that of the germanium bolometer, but the time constant is 10^{-7} sec, considerably shorter.

The photoconductive effect occurring in semiconducting materials provides a class of quantum detectors that are particularly useful in the near infrared. The incident photon flux interacts with electrons in bound states in the valence band or trapping level and excites them into free states in the conduction band, where they remain for a characteristic lifetime. A positive hole is left behind. Both electron and hole contribute to electrical conduction. At wavelengths longer than about 10 μm, photons have insufficient energy to promote electrons across the forbidden zone between the valence and conduction bands. Semiconductors must be doped with impurity atoms to provide some intermediate energy levels. Lead sulfide detectors are sensitive to radiation below 3 μm, lead telluride below 4 μm, and lead selenide below 5 μm. These longwavelength limits are increased about 50% on cooling detectors to 20 °K. Doped germanium and silicon detectors cooled to 4 °K possess useful sensitivity at wavelengths as long as 120 μm. Response time for lead selenide and telluride is less than 10 μsec and for doped germanium and silicon, about 1 nsec. The speed of response is determined by the time required for the excited charge carriers to become immobilized through recombination. These detectors consist of a film of semiconductor 0.1-μm thick deposited on a glass or quartz base and sealed into an evacuated envelope. The limiting noise is generation recombination noise associated with fluctuations in the density of free-charge carriers; it is produced either by lattice vibrations, when the detector is not cooled sufficiently or the random arrival of photons from the background.

The pyroelectric detector consists of a slice of a ferroelectric material, usually a single crystal of triglycine sulfate. Radiation absorbed by the crystal is converted into heat, which alters the lattice spacings. Below the Curie temperature of the crystal, a change in lattice spacings produces a change in spontaneous electric polarization. The crystal is mounted between two parallel electrodes, one of them infrared transparent and both normal to the polarization axis. If the electrodes are connected to an external circuit, a current is set up to balance the polarization charge at the crystal faces. This current then produces a voltage signal across a appropriate load resistor, which is applied directly to a field-effect transistor, an integral part of the detector package. A distinct advantage of this detector is the decrease of its sensitivity (close to that of a Golay detector) by $\omega^{-1/2}$ when the chopping frequency ω of the beam increases instead of the ω^{-1} relation in most other thermal detectors.

8.1.3.2. Spectrometers

Most infrared spectrophotometers are double-beam instruments in which two equivalent beams of radiant energy are taken from the source. By means of a combined rotating mirror and light interrupter, the source is alternately flicked between the reference and sample paths. In the optical-null system, the detector responds only when the intensity of the two beams is unequal. Any imbalance is corrected for by a light attenuator (an optical wedge or shutter comb) moving in or out of the reference beam to restore balance. The recording pen is attached to the light attenuator. Although very popular, the optical-null system has serious faults. Near zero transmittance of the sample, the reference beam attenuator stops practically all light in the reference beam. Both beams are then blocked, no energy passes, the spectrometer has no way of determining how close it is to the correct transmittance value, and the instrument stops functioning. However, in the midinfrared region, the electrical beam-ratioing method is not an easy means of avoiding the deficiencies of the optical-null system. To a large extent it is trading optical and mechanical problems for electronic problems.

Monochromators employing prisms for dispersion use a Littrow 60° prism plane mirror mount. Midinfrared instruments employ a sodium chloride prism for the region from 4000–650 cm^{-1} (2.5–15.4 μm), a potassium bromide or cesium iodide prism and optics extend the useful spectrum to 400 cm^{-1} (25 μm) or 270 cm^{-1} (37 μm), respectively. Quartz monochromators, designed for the ultraviolet visible region, extend their coverage into the near infrared to 2500 cm^{-1} (4 μm).

Plane-reflectance-grating monochromators dominate today's instruments. To cover the wide wavelength range, several gratings with different ruling densities and associated higher order filters are necessary. This requires some complex sensing and switching mechanisms to automate the scan with acceptable accuracy. Because of the nature of the blackbody emission curve, a slit-programming mechanism must be employed to give near constant energy and resolution as a function of wavelength. The principal limitation is energy. Resolution and signal-to-noise ratio are limited primarily by the emission of the blackbody source and the noise equivalent power of the detector. Two gratings are often mounted back-to-back so that each is used only in the first order, and the gratings are changed to 2000 cm^{-1} (5 μm) in midinfrared spectrometers. Grating instruments incorporate a sine bar mechanism to drive the grating mount when a wavelength readout is desired and a cosecant bar drive when wavenumbers are desired. Undesired overlapping can be eliminated with a fore prism or suitable filters.

The filters are inserted near a slit or slit image when the required size of the filter is not excessive. The circular variable filter is simple in construction. It is frequently necessary to use gratings as reflectance filters when working in the far infrared to remove unwanted second and higher orders from the light incident on the far-infrared grating. For this purpose, small plane gratings blazed for the wavelength of the unwanted radiation are used. The grating acts as a mirror, reflecting the desired light into the instrument and diffracting the shorter wavelengths out of the beam; in fact, grating "looks" like a good mirror to wavelengths longer than the groove spacing.

Probably the most elegant filter is a prism, because it provides a narrow band of wavelengths with high efficiency over a relatively broad spectral range. The prism and grating must track together over consecutive grating orders. Light from the parabolic mirror enters the fore prism where it is dispersed so that only a relatively narrow band of wavelengths is allowed to fall on the grating. The resolution of the prism can be quite low, because it need only exclude the adjacent orders, but for the higher orders, the interval becomes successively narrower. Thus, it is preferable to use two gratings and confine their application to lower orders.

8.1.3.3. Dispersive and Nondispersive

A dispersive instrument is patterned after a double-beam spectro-photometer. Radiation at two fixed wavelengths passes through a cell containing the process stream to provide a continuous measurement of the absorption ratio. At one wavelength, the material absorbs selectively; at the other wavelength, the material does not absorb or else exhibits a constant but small absorption. The ratio of transmittance readings is converted directly into concentration of absorber and recorded. This type of instrument can handle liquid systems as well as gas streams, and it is able to analyze quite complex mixtures.

No prisms or gratings are used in nondispersive instruments. The total radiation from an infrared source is passed through the sample, providing more signal power. By filling one or both cells containing the detector with the pure form of the gas being determined, these analyzers show high selectivity and virtually infinite resolving power, although they employ a very simple optical train. Two modes of detection are employed.

In the negative filter type of nondispersive analyzer, an infrared source sends radiation through the sample chamber. Half of the beam is intercepted by each detector. One cell is filled with the pure form of the gas being determined (component A); the other is filled with a nonabsorbing gas. The former absorbs all the radiation in its beam that is characteristic of component A, and a thermal detector in the cell records the temperature rise. As radiation passes through a gas stream in the sample cell devoid of component A, the detector filled with the nonabsorbing gas absorbs some of the radiant energy. When the process stream contains some of component A, a proportionate amount of the characteristic radiation is absorbed in the sample cell and fails to reach the detectors, thus decreasing the signal from the detector filled with pure component A, no change occurs in the signal from the other detector. Thus, as the concentration of component A in the process stream approaches 100%, the signal difference between the two detectors approaches zero.

In the positive filter arrangement, the beam of radiation from the source is split into two parallel beams. One beam passes through the reference cell and the other through the sample cell. In this arrangement, each detector is filled with the pure form of the gas being determined. When some of the latter is present in the sample beam, the sample detector receives less radiant energy by the amount absorbed by the sample component at its characteristic wavelength. The signal difference between the two detectors is related to absorber concentration.

If some other absorbing component is present in the process stream, the analyzer is "desensitized" by filling an intermediate cell (in both light paths) with the pure form of the interfering gas or a sufficient concentration to remove its characteristic wavelengths adequately from both light paths. The analyzer operates on the remaining regions on the spectrum. Of course, this somewhat reduces the sensitivity of the analyzer toward the component being analyzed.

8.1.4. Interferometric (Fourier Transform) Spectrometer

The basic configuration of the interferometer portion of a Fourier transform spectrometer includes two plane mirrors at a right angle to each other and a beam splitter at 45° to the mirrors (Low, 1970). Modulated light from the source is collimated and passes to the beam splitter, which divided it into two equal beams for the two mirrors. An equal thickness of support material (without the semireflection coating), called the compensator, is placed in one arm of the interferometer to equalize the optical path length in both arms. When these mirrors are positioned so that the optical path lengths of the reflected and transmitted beams are equal, the two beams are in phase when they return to the beam splitter and constructively interfere. Displacing the movable mirror by one-quarter wavelength moves the two beams 180° out of phase, causing them to interfere destructively. Continuing to move the mirror in either direction causes the field to oscillate from light to dark for each quarter-wavelength movement of the mirror, corresponding to $\lambda/2$ changes. When the interferometer is illuminated by monochromatic light of wavelength λ and the mirror is moved with a velocity v, the signal from the detector has a frequency $f = 2v/\lambda$. A plot of signal versus mirror distance is a pure cosine wave. With polychromatic light, the output signal is the sum of all the cosine waves, which is the Fourier transform of the spectrum. Each frequency is given an intensity modulation f, which is proportional both to the frequency of the incident radiation and to the speed of the moving mirror. For example, with a constant mirror velocity of 0.5 mm/sec, radiation of 1000 cm^{-1} (10 μm and a frequency of 3×10^{14} Hz) produces a detector signal of 50 Hz. For 5-μm radiation, the signal is 100 Hz, *etc*. An appropriate inverse transformation of the interferogram gives the desired spectrum. Rather than dispersing polychromatic radiation like a conventional dispersive spectrometer, the Fourier transform spectrometer performs a frequency transformation. Data reduction requires digital computer techniques and analog conversion devices.

To interpret the intensity measurement, the displacement of the movable mirror has to be known precisely. With a constant velocity of mirror motion, the mirror should move as far and as smoothly as possible. If the velocity is precise, an electronically timed coordinate can be generated for the interferogram. Severe mechanical problems limit this approach. The interferometer itself, however, can be used to generate its own time scale. In addition to processing incoming spectral radiation, a line from a laser source produces a discrete signal that is time-locked to the mirror motion and hence to the interferogram. This is the fringe reference system, which is analogous to the frequency/field lock in nuclear magnetic resonance (nmr). The mirror position can be determined by measuring the laser line interferogram, counting the fringes as the mirror moves from the starting position—these are indicated by a burst of light from an incandescent source.

Dispersion or filteration is not required, so that energy-wasting slits are not needed, which is a major advantage. With energy at a premium in the far infrared, the superior light-gathering power of the interferometric spectrometer is an asset.

In the near-infrared and midinfrared, fermanium coating on a transparent salt, such as NaCl, KBr, or CsI, is a common beam splitter material. In far-infrared spectrometers, the beam splitter is a thin film of Mylar whose thickness must be chosen for the spectral region of interest. For example, a Mylar film 0.25 mil thick can effectively cover the range from 60–375 cm^{-1}.

Resolution is related to the maximum extent of mirror movement, so that a 1-cm movement results in 1-cm^{-1} resolution, and a 2-cm movement yields 0.1-cm^{-1} resolution. Resolution can be doubled by doubling the measurement times, or resolution can be traded for rapid response. Because the detector of the interferometer sees all resolution elements throughout the entire scan time, the signal-to-noise ratio S/N, is proportional to T, where T is the measurement time. For example, when examining a spectrum composed of 2000 resolution elements with an observation time of 1 sec per element assumed for the desired S/N, the interferometric measurement is complete in 1 sec. Improving the S/N by a factor of 2 would require only 4 sec to complete the measurement. Comparable times for a dispersive spectrometer are 33 and 72 min, respectively. Repetitive signal-averaged scans are very feasible with an interferometer.

8.1.5. Sampling Handling

Infrared instrumentation has reached a remarkable degree of standardization as far as the sample compartment of various spectrometers is concerned. Sample handling itself, however, presents a number of problems in the infrared region. No rugged window material for cuvettes exists that is transparent and also inert over this region. The alkali halides are widely used, particularly sodium chloride, which is transparent at wavelengths as long as 16 μm (625 cm^{-1}); however cell windows are easily fogged by exposure to moisture and require frequent repolishing. Silver chloride is often used for moist samples, or aqueous solutions, but it is soft, easily deformed, and darkens on exposure to visible light. Teflon has only C—C and C—F absorption band. For frequencies under 600 cm^{-1}, a polyethylene cell is useful. Crystals with a high refractive index produce strong, persistent fringes.

8.1.5.1. Liquids and Solutions

Samples that are liquid at room temperature are usually scanned in their next form or in solution. The sample concentration and path length should be chosen so that the transmittance lies between 10 and 70%. For neat liquids, this represents a very thin layer, about 0.001–0.05 mm in thickness. For solutions, concentrations of 10% and cell lengths of 0.1 mm are most practical. Unfortunately, not all substances can be dissolved in a reasonable concentration in a solvent that is nonabsorbing in regions of interest. When possible, the spectrum is obtained in a 10% solution of CCl_4 in a 0.1-mm cell in the region 4000–1333 cm^{-1} (2.5–7.5 μm) and in a 10% solution of CS_2 in the region 1333–650 cm^{-1} (7.5–15.4 μm). Chloroform, methylene chloride, acetonitrile, and acetone are used to obtain solution spectra of polar materials insoluble in CCl_4 or CS_2. Sensitivity can be gained by going to longer path lengths if a suitably transparent solvent can be found. In a double-beam spectrophotometer, a reference cell of the same path length as the sample cell is filled with pure solvent and placed in the reference beam. Moderate solvent absorption, now common to both beams, is not observed in the recorded spectrum; however, solvent transmittance should never fall below 10%.

The possible influence of a solvent on the spectrum of a solute must not be overlooked. Particular care must be exercised in selecting a solvent for compounds susceptible to hydrogen-bonding effects. Hydrogen bonding through an —OH or —NH— group alters the characteristic vibrational frequency of that group: The stronger the hydrogen bonding, the greater the fundamental frequency is lowered. To differentiate between inter- and intramolecular hydrogen bonding, a series of spectra at different dilutions but with the same number of absorbing molecules in the beam must be obtained. If as dilution increases, the hydrogen-bonded absorption band decreases while the unbonded absorption band increases, the bonding is intermolecular. Intramolecular bonding shows no comparable dilution effect.

Infrared solution cells are constructed with windows sealed and separated by thin gaskets of copper and lead wetted with mercury. The whole assembly is securely clamped together. As the mercury penetrates the metal, the gasket expands, producing a tight seal. The cell is provided with tapered fittings to accept hypodermic syringe needles for filling. In demountable cells, the sample and spacer are placed on one window, covered by another window, then the entire "sandwich" is clamped together.

8.1.5.2. Cell Thickness

One of two methods can be used to measure the path length of infrared absorption cells: the interference fringe method or the standard absorber method. The interference fringe method is ideally suited to cells whose windows have a high polish. With the empty cell in the spectrophotometer on the sample side and no cell in the reference beam, the spectrophotometer is operated as near as possible to the 100% line. Enough spectrum is run to produce 20–50 fringes. The cell thickness b (in centimeters) is calculated from the expression

$$b = \frac{1}{2\eta_D}\left(\frac{n}{\nu_1 - \nu_2}\right)$$

where n is the number of fringes (peaks or troughs) between two wavenumbers ν_1 and ν_2, and η_D is the refractive index of the sample material. If measurements are made in wavelength (micrometers), the equation is

$$b = \frac{1}{2\eta_D}\left(\frac{n\lambda_1\lambda_2}{\lambda_2 - \lambda_1}\right)$$

where λ_1 is the starting wavelength and λ_2 the finishing wavelength. The fringe method also works well for measuring film thickness.

The standard absorber method can be used with a cell in any condition on cavity cells whose inner faces do not have a polished finish. The 1960-cm^{-1} (5–10-μm) band of benzene can be used for calibrating cells less than 0.1 mm in path length, and the 845-cm^{-1} (11.8-μm) and for cells 0.1 mm or longer in path length. At the former frequency, benzene has an absorbance of 0.10 for every 0.01 mm of thickness; at 845 cm^{-1}, benzene has an absorbance of 0.24 for every 0.1 mm of thickness.

8.1.5.3. Films

Spectra of liquids insoluble in a suitable solvent are best obtained from capillary films. A large drop of the neat liquid is placed between two rock salt plates that are then squeezed together and mounted in the spectrometer in a suitable holder. Plates need not have high polish, but they must be flat to avoid distorting the spectrum.

For polymers, resins, and amorphous solids, the sample is dissolved in any reasonably volatile solvent, the solution poured onto a rock salt plate, and the solvent evaporated by gentle heating. If the solid is noncrystalline, a thin homogeneous film is deposited on the plate, which then can be mounted and scanned directly. Sometimes polymers can be "hot pressed" onto plates.

8.1.5.4. Mulls

Powders, or solid reduced to particles, can be examined as a thin paste or mull by grinding the pulverized solid (about 9 mg) in a greasy liquid medium. The suspension is pressed into an annular groove in a demountable cell. Multiple reflections and reflections from the particles are reduced by grinding particles to a size an order of magnitude less than the analytic wavelength and by surrounding them in a medium whose refractive index more closely matches theirs than does air. Liquid media include mineral oil or Nujol, hexachlorobutadiene, perfluorokerosene, and chlorofluorocarbon gases (fluorolubes). The latter are used when absorption by the mineral oil masks the presence of C—H bands. For qualitative analysis, the mull technique is rapid and convenient, but quantitative data are difficult to obtain even when an internal standard is incorporated into the mull. Polymorphic changes, degradation, and other changes may occur during grinding.

8.1.5.5. Pellet Technique

The pellet technique involves mixing a finely ground sample (1–100 μg) and potassium bromide powder, then pressing the mixture into an evacuable die at sufficient pressure (60,000–100,000 psi) to produce a transparent disk. Grinding and mixing are conveniently done in a vibrating ball mill (Wig-L-Bug). Other alkali halides may also be used, particularly CsI or CsBr, for measurements at longer wavelengths. Good dispersion of the sample in the matrix is critical, and there cannot be any moisture; freeze-drying the sample is often a necessary preliminary step.

The KBr wafers can be formed without evacuation in a Mini-Press®, where two highly polished bolts turn against one another in a steel cylinder. Pressure is applied with wrenches for about 1 min to 75–100 mg of powder, then the bolts are removed, and the cylinder—now a cell complete with window—is installed in its slide holder in any spectrophotometer. Quantitative analyses can be performed, since the weight ratio of a sample to the internal standard added in each disk or wafer can be measured.

8.1.6. Attenuated Total Reflectance

The scope and versatility of infrared spectroscopy as a qualitative analytic tool have been increased substantially by attenuated total reflectance (ATR), also known as internal reflectance technique (Harrick, 1967; Wilkes, 1972). When a beam of radiation enters a plate (or prism), it is reflected internally if the angle of incidence at the interface between sample and plate is greater than the critical angle (which is a function of refractive index). Or internal reflection, all the energy is reflected; however, the beam appears to penetrate slightly (from a fraction of a wavelength up to several wavelengths) beyond the reflecting surface, and then return. When a material is placed in contact with the reflecting surface, the beam loses energy at those wavelengths where the material absorbs due to an interaction with the penetrating beam. This attenuated radiation, when measured and plotted as a function of wavelength, gives rise to an absorption spectrum characteristic of the material that resembles an infrared spectrum obtained in the normal manner.

Most ATR work is done by means of an accessory readily inserted in, and removed from, the sampling space of a conventional infrared spectrophotometer. The accessory consists of a mirror system that sends source radiation through the attachment and a second mirror system that directs the radiation into the monochromator. The width of the crystal is chosen to be equal to, or greater than, the height of the spectrometer slits (10–15 mm). The length-to-thickness ratio determines the number of reflections once the angle of incidence is selected; dimensions vary from 0.25–5 mm of thickness and lengths from 1–10 cm. Parallelism and flatness of sampling surfaces and surface polish are critical. In the single-pass plate, light is introduced through an entrance aperture consisting of a simple bevel at one end of the plate and after propagation via multiple internal reflection down the length of the plate, leaves by means of an exit aperture either parallel or perpendicular to the entrance aperture. The angle of the bevel determines the interior angle of incidence. This type of plate is useful for bulk materials, thin films, and surface studies. In the double-pass plate, light enters as before, propagates down the length of the plate, is totally reflected at the opposite end from a surface perpendicular to the sample surfaces, then returns to leave the plate via the exit aperture. The free end of the plate can be dipped into liquids or powders or placed in closed systems requiring only one optic window.

The apparent depth to which radiation penetrates the sample is only a few micrometers, and it is independent of sample thickness. Consequently, ATR spectra can be obtained for many samples that cannot be studied by normal transmission methods. These include samples that show very strong absorptions, resist preparation in thin films, and are available on a nontransparent support. Aqueous solutions can be handled without compensating for very strong solvent absorptions. Samples containing suspended matter, such as dispersed solids or emulsions, that produce high backgrounds in emission spectra due to scatter yield better results with ATR. Excellent contact efficiency at the sample and crystal interface is achieved when then sample is a self-adhering mobile liquid, flows slightly under modest pressure, or can be evaporated from solution. Samples that absorb only weakly or do not have intimate contact with the crystal surface, such as rough-textured fabric, can be handled using multiple internal reflections, analogous to increasing the path length in transmission studies.

The appearance and intensity of an ATR spectrum depends on the difference between the refraction indices for reflection crystal and the rarer medium containing the absorber, and on the internal angle of incidence. Thus, a reflection crystal with a relatively high refraction index should be used. Two materials found to perform most satisfactorily for the majority of liquid and solid samples are KRS-5 and AgCl. The KRS-5 is a tough and durable material with excellent transmission properties. Its refraction index is high enough to permit well-defined spectra of nearly all organic materials, although it is soluble in basic solutions. The AgCl is recommended for aqueous samples because of its insolubility and lower refraction index. An overall angle of incidence should be selected that is far enough from the average critical angle of sample versus reflector so that the change in the critical angle through the region of changing refraction index (the absorption band) has a minimum effect on the shape of the ATR band. Unfortunately, when the refraction index for the crystal is considerably greater than that for the sample, so that little distortion occurs, total absorption is reduced. With multiple-reflection equipment, however, ample absorption can be obtained at angles well away from the critical angle.

8.1.7. Infrared Probe

The infrared probe, which resembles a specific ion electrode, contains a sensitive element that is dipped into the sample. To operate the probe, (1) the user selects the proper wavelength by rotating a calibrated, circular variable filter; (2) then adjusts the gain and slits to bring the meter to 100%; (3) next, the probe is lowered into the sample. The meter indicates the absorbance. This value can be converted into concentration by reference to a previously prepared calibration curve. To detect the presence or absence of a particular functional group, the user scans through the portion of the spectrum where absorption bands characteristic of that group appear.

The infrared probe uses ATR to obtain the absorption information. The probe crystal is made from a chemically inert material, such as germanium or synthetic sapphire. The reflecting surfaces are masked so that the same area is covered by a sample each time an analysis is made. A single-beam optic system is employed, chopped at 45 Hz. Since the air path is less than 5 cm, as opposed to well over 1 m in conventional infrared spectrophotometers, absorption due to atmospheric water vapor and carbon dioxide is insignificant.

8.1.8. Quantitative Analysis

Infrared spectroscopy as a quantitative analytic tool varies widely from one laboratory to another, but high-resolution grating instruments considerably increase the scope and reliability of quantitative infrared analysis, which is based on Beer's law. Apparent deviations in analysis results arise from either chemical or instrumental effects. In many cases, the presence of scattered radiation makes direct application of Beer's law inaccurate, especially at high values of absorbance. Since the energy available in the useful portion of the infrared is usually quite small, it is necessary to use rather wide slit widths in the monochromator. This causes a considerable change in the apparent value of the molar absorptivity, which should therefore be determined empirically.

The baseline method of quantitative analysis involves selecting an absorption band of the substance under analysis that does not fall too close to the bands of other matrix components. The value of the incident radiant energy P_0 is obtained by drawing a straight line tangent to the spectral absorption curve at the position of the sample's absorption band. The transmittance P is measured at the point of maximum absorption. The value of $\log(P_0/P)$ is then plotted against concentration.

Many possible errors are eliminated by the baseline method because the same cell is used for all determinations, all measurements are made at points on the spectrum sharply defined by the spectrum itself, so there is no dependence on wavelength settings. Using such ratios eliminates changes in instrument sensitivity, source intensity, or changes in adjustment of the optic system.

Pellets from the disk technique can be employed in quantitative measurements; uniform pellets of similar weight are essential, however, for quantitative analysis. Known weights of KBr are used and a known quantity of the test substance from which absorbance data and a calibration curve can be constructed. The disks are weighted and their thickness measured at several points on the surface with a dial micrometer. The disadvantage of measuring pellet thickness can be overcome by using an internal standard, such as potassium thiocyanate, which should be preground, dried, and then reground at a concentration of 0.2% by weight with dry KBr. The final mix is stored over phosphorous pentoxide. A standard calibration curve is made by mixing about 10% by weight of the test substance with the KBr–KSCN mixture and then grinding the mixture to a fine powder. Measurement of the thiocyanate absorption at 2125 cm^{-1} (4.70 μm) to a chosen absorption of the test substance is plotted against percent concentration of the sample.

For quantitative measurements, the single-beam system has some fundamental characteristics that can result in greater sensitivity and better accuracy than double-beam systems. All other things being equal, a single-beam instrument will automatically have a greater signal-to-noise ratio: There is a factor of 2 advantage in looking at one beam all the time rather than two beams half the time. Electronic switching gives another factor of 2 advantage. Thus, in any analytic situation where background noise is appreciable, the single-beam spectrometer should be superior.

8.1.9. Correlating Infrared Spectra with Molecular Structure

The infrared spectrum of a compound is essentially the superposition of absorption bands of specific functional groups, yet subtle interactions with the surrounding atoms of the molecule impose the stamp of individuality on the spectrum of each compound. For qualitative analysis, one of the best features of an infrared spectrum is that absorption or lack of absorption in specific frequency regions can be correlated with specific stretching and bending motions and in some cases, with the relationship of these groups to the remainder of the molecule. Thus by interpretating the spectrum, it is possible to state that certain functional groups are present in the material and certain others are absent. With this one datum, the identity of the unknown can sometimes be so sharply narrowed that comparison with a library of pure spectra permits identification.

8.1.9.1. Near-Infrared Region

In the near-infrared region, which meets the visible region at about 12,500 cm^{-1} (0.8 μm) and extends to about 4000 cm^{-1} (2.5 μm), there are many absorption bands resulting from harmonic overtones of fundamental bands and combination bands often associated with hydrogen atoms. Among these are the first overtones of the O—H and N—H stretching vibrations near 7140 cm^{-1} (1.4 μm) and 6667 cm^{-1} (1.5 μm), respectively; combination bands resulting from C—H stretching; and deformation vibrations of alkyl groups at 4548 cm^{-1} (2.2 μm). Thicker sample layers (0.5–10 mm) compensate for reduced molar absorption. The near-infrared region is accessible with quartz optics, and this is coupled with greater sensitivity of near-infrared detectors and more intense light sources. The near-infrared region is often used for quantitative work.

Water has been analyzed in glycerol, hydrazine, Freon, organic films, acetone, and fuming nitric acid. Absorption bands at 2.76, 1.90, and 1.40 μm are used, depending on the concentration of the test substance. When interferences from other absorption bands are severe or very low concentrations of water are being studied, water can be extracted with glycerol or ethylene glycol.

Near-infrared spectrometry is a valuable tool for analyzing mixtures of aromatic amines. Primary aromatic amines are characterized by two relatively intense absorption bands near 1.97 and 1.49 μm. The band at 1.97 μm is a combination of N—H bending and stretching modes, and the one at 1.49 μm is the first overtone of the symmetric N—H stretching vibration. Secondary amines exhibit an overtone band, but they do not absorb appreciably in the combination region. Secondary amines exhibit an overtone band, but they do not absorb appreciably in the combination region. These differences in absorption provide the basis for rapid quantitative analytic methods. Analyses are normally carried out on 1% solutions in CCl_4, using 10-cm cells. Background corrections can be obtained at 1.575 and 1.915 μm. Tertiary amines do not exhibit appreciable absorption at either wavelength. The overtone and combination bands of aliphatic amines are shifted to about 1.525 and 2,000 μm, respectively. Interference from the first overtone of the O—H stretching vibration at 1.40 μm is easily avoided with the high resolution available with near-infrared instruments.

8.1.9.2. Midinfrared Region

Many useful correlations have been found in the midinfrared region, which is divided into the "group frequency" region of 4000–1300 cm^{-1} (2.5–8 μm) and the "fingerprint" region of 1300–650 cm^{-1} (8.0–15.4 μm). In the group frequency region, the principal absorption bands can be assigned to vibration units consisting of only two atoms of a molecule, that is, units more or less dependent only on the functional group giving the absorption, and not on the complete molecular structure. Structural influences do reveal themselves, however, as significant shifts from one compound to another. In the deviation of information from an infrared spectrum, prominent bands in this region are noted and assigned first. In the interval from 4000–2500 cm^{-1} (2.5–4 μm), absorption is characteristic of hydrogen-stretching vibrations with elements of mass 19 or less. When coupled with heavier masses, the frequencies overlap the triple-bond region. The intermediate frequency range from 2500–1540 cm^{-1} (4–6.5 μm), is often termed the unsaturated region. Triple bonds, and very little else, appear from 2500–2000 cm^{-1} (4–5 μm). Double-bond frequencies fall in the region from 2000–1540 cm^{-1} (5–6.5 μm). By judicious application of accumulated empirical data, it is possible to distinguish among C=O, C=C, C=N, N=O, and S=O bands. The major factors in the spectra between 1300–650 cm^{-1} (7.7–15.4 μm) are single-band stretching frequencies and bending vibrations (skeletal frequencies) of polyatomic systems, which involve motions of bonds linking a substituent group of the remainder of the molecule. This is the fingerprint region. Multiplicity is too great for assured individual identification, but collectively the absorption bands aid in identification.

8.1.9.3. Far-Infrared Region

The region between 667 and 10 cm^{-1} (15–1000 μm) contains the bending vibrations of carbon, nitrogen, oxygen, fluoride with atoms heavier than mass 19, and additional bending motions in cyclic or unsaturated systems. The low-frequency molecular vibrations found in the far infrared are particularly sensitive to changes in the overall structure of the molecule. When studying the conformation of the molecule as a whole, the far-infrared bands often differ in a predictable manner for different isometric forms of the same basic compound. The far-infrared frequencies of organometallic compounds are often sensitive to the metal ion or atom, and this, too, can be used advantageously in studying coordination bonds. Moreover, this region is particularly well suited to studying organometallic or inorganic compounds whose atoms are heavy but whose bonds are inclined to be weak (Ferraro, 1968).

8.1.9.4. Structural Analysis

After the presence of a particular fundamental stretching frequency has been established, closer examination of the shape and exact position of an absorption band often yields additional information. The shape of an absorption band around 3000 cm^{-1} (3.3 μm) gives a rough idea of the CH group present. Alkyl groups have their C—H stretching frequencies lower than 3000 cm^{-1}, whereas they are slightly higher than 3000 cm^{-1} for alkenes and aromatics. The CH$_3$ group gives rise to an asymmetric stretching mode at 2960 cm^{-1} (3.38 μm) and a symmetric mode at 2870 cm^{-1} (3.48 μm), for —CH$_2$—, these bonds occur at 2930 cm^{-1} (3.42 μm) and 2850 cm^{-1} (3.51 μm).

Next, it is important to examine regions where characteristic vibrations from bending motions occur. For alkanes, bands at 1460 cm^{-1} (6.85 μm) and 1380 cm^{-1} (7.25 μm) are indicative of a terminal methyl group attached to carbon-exhibiting in-plane bending motions; if the latter band is split into a doublet at about 1397 cm^{-1} and 1370 cm^{-1} (7.16 μm and 7.30 μm, respectively), this indicates geminal methyls. The symmetrical in-plane bending is shifted to lower frequencies when the methyl group is adjacent to >C=O (1360–1350 cm^{-1}), —S— (1325 cm^{-1}), and silicon (1250 cm^{-1}), the in-plane scissor motion of —CH$_2$— at 1470 cm^{-1} (6.80 μm) indicates the presence of that group. Four or more methylene groups in a linear arrangement gives rise to a weak rocking motion at about 720 cm^{-1} (13.9 μm).

The substitution pattern of an aromatic ring can be deduced from a series of weak but very useful bands in the region of 2000–1670 cm^{-1} (5–6 μm) coupled with the position of strong bands between 900–650 cm^{-1} (11.1–5.4 μm), which are due to the out-of-plane bending vibrations. Absence of the asymmetrical breathing mode at 690–710 cm^{-1} (14.1–14.5 μm) in the spectra of para- and ortho-substituted rings are helpful. Ring stretching modes are observed near 1600 cm^{-1}, 1570 cm^{-1}, and 1500 cm^{-1} (6.25 μm, 6.37 μm, and 6.67 μm, respectively). These characteristic absorption patterns are also observed with substituted pyridines and polycyclic benzenoid aromatics.

The presence of an unsaturated C=C linkage introduces the stretching frequency at 1650 cm^{-1} (6.07 μm), which may be weak or nonexistent if symmetrically located in the molecule. Mono- and tri-substituted olefins give rise to more intense bands that *cis-* or *trans-*distributed olefins. Substitution by a nitrogen or oxygen functional group greatly increases the intensity of the C=C absorption band. Conjugation with an aromatic nucleus causes a slight shift to lower frequency, but with a second C=C or C=O, the shift to lower frequency is 40–60 cm^{-1} (250.0–166.7 μm), with a substantial increase in intensity. The out-of-plane bending vibrations of the hydrogens on a C=C linkage are very valuable. A vinyl group gives rise to two bands at about 990 cm^{-1} (10.1 μm) and 910 cm^{-1} (11.0 μm). The =CH$_2$ (vinylidene) band appears near 895 cm^{-1} (11.2 μm), and it is a very prominent feature of the spectrum. *Cis-* and *trans-*disubstituted olefins absorb near 685–730 cm^{-1} (13.7–14.6 μm) and 965 cm^{-1} (10.4 μm), respectively. The single hydrogen in a tri substituted olefin appears near 820 cm^{-1} (12.2 μm).

In alkynes, the ethynyl hydrogen appears as a needle-sharp and intense band at 3300 cm^{-1} (3.0 μm). The absorption band for —C≡C— is located in about the range from 2100–2140 cm^{-1} (4.67–4.76 μm) when terminal but in the range from 2260–2190 cm^{-1} (4.42–4.56 μm) if nonterminal. The intensity of the latter type band decreases as symmetry of the molecule increases. When the acetylene linkage is conjugated with a carbonyl group, however, absorption becomes very intense. The molecule is best identified by Raman spectroscopy.

For ethers, the one important band appears near 1100 cm^{-1} (9.09 μm); it is due to the antisymmetric stretching mode of the —C—O—C— links. Since this band is quite strong, it may dominate the spectrum of a simple ether.

For alcohols, the most useful absorption is due to stretching of the O—H bond. In the free or unassociated state, the band appears weak but sharp at about 3600 cm^{-1} (2.78 μm). Hydrogen bonding greatly increases the intensity of the band and moves it to lower frequencies; if hydrogen bonding is especially strong, the band becomes quite broad. Intermolecular hydrogen bonding is concentration dependent, whereas intramolecular hydrogen bonding is not. Measurements in solution under different concentrations are invaluable. The spectrum of an acid is quite distinctive in shape and breadth in the high-frequency region. The distinction between the several types of alcohol is often possible on the basis of the C—O stretching absorption bands.

The carbonyl group is not difficult to recognize, since it is often the strongest band in the spectrum. Its exact position in the region, extending from about 1825–1575 cm^{-1} (5.48–6.35 μm) depends on the double-bond character of the carbonyl group. Anhydrides usually show a double absorption band. Aldehydes are distinguished from ketones by an additional C—H stretching frequency in the CHO group at about 2720 cm^{-1} (3.68 μm). In esters, two bands related to C—O stretching and bending are recognizable between 1300 and 1040 cm^{-1} (7.7–9.6 μm), in addition to the carbonyl band. The carboxyl group, in a sense, shows bands arising from the superposition of C=O, C—O,

C—OH, and O—H vibrations. Of five characteristic bands, three of these (2700 cm^{-1}, 1300 cm^{-1} and 943 cm^{-1} [3.7 μm, 7.7 μm, and 10.6 μm]) are associated with vibrations in the carboxyl OH, which disappear when the carboxylate ion is formed. When the acid exists in the dimeric form, the O—H stretching band at 2700 cm^{-1} (3.7 μm) disappears, but the absorption band at 943 cm^{-1} (10.6 μm) due to OH out-of-plane bending of the dimer remains.

Of particular interest in a primary amine (or amide) are the N—H stretching vibrations at about 3500 cm^{-1} and 3400 cm^{-1} (2.86 μm and 2.94 μm), the in-plane bending of N—H at 1610 cm^{-1} (6.2 μm), and the out-of-plane bending of —NH$_2$ at about 830 cm^{-1} (12.0 μm), which is broad for primary amines. By contrast, a secondary amine exhibits a single band in the high-frequency region at about 3350 cm^{-1} (2.98 μm). The high-frequency bands broaden and shift about 100 cm^{-1} (100 μm) to lower frequency when involved in hydrogen bonding. When the amine salt is formed, these bands are markedly broadened and lie between 3030 and 2500 cm^{-1} (3.3 and 4.0 μm), resembling COOH bands in this region.

The nitro group is characterized by two equally strong absorption bands at about 1560 cm^{-1} and 1350 cm^{-1} (6.41 μm and 7.40 μm), the asymmetric and symmetric stretching frequencies. In an N-oxide, only a single very intense band is present in the region from 1300–1200 cm^{-1} (7.70–8.33 μm). In addition, there are C—N stretching and various bending vibrations whose positions should be checked. Quite analogous bands are observed for bonds between S and O; all are intense. Stretching frequencies of SO$_2$ appear around 1400–1310 cm^{-1} and 1230–1120 cm^{-1} (7.14–7.63 μm and 8.13–8.93 μm); for S=O at 1200–1040 cm^{-1} (8.33–9.62 μm); and for S—O around 900–700 cm^{-1} (11.11–14.28 μm).

8.1.10. Identifying Compounds

In many cases, interpreting the infrared spectrum on the basis of characteristic frequencies does not permit positive identification of an unknown but only the type of compound class. The tendency to overinterpret a spectrum, that is, interpret and assign all of the observed absorption bands, particularly those of moderate and weak intensity in the fingerprint region, must be resisted. Once the category is established, the spectrum of the unknown is compared with spectra of appropriate known compounds for an exact spectral match. If the exact compound does not happen to be in the file, particular structural variations within the category may be helpful in suggesting possible answers and eliminating others. Several collections of spectra are commercially available (*ASTM-Wynadotte Index*, 1963; Nyquist *et al.*, 1971; Putley, 1970; Sadtler Research Laboratories, 1963; Pouchert, 1970).

Lead-based paint consists of a limited number of organic materials (primarily resins and pigments), which can be analyzed by infrared spectroscopy. Appendix B contains a list of paint binders, and additional spectra for coatings are found in Afremow (1969). The reader should review material in "Fundamentals of Paint" to correlate paint compounds with their corresponding infrared spectra. An excellent text to consult on performing and interpreting infrared spectra analysis is Willard *et al.* (1974).

8.2. THERMAL ANALYSIS

8.2.1. Thermogravimetric Analysis

The total amount of pigment (percent solids) in a paint film can be determined by plotting percent weight versus temperature, using a thermogravimetric analyzer because temperature varies (*e.g.*, 20°C/min or 68°F/min) to decompose the organic (binder) portion of the film and leave the thermally stable pigments, assuming inorganic pigments. The decomposition temperature also indicates the chemical structure of the binders, because chemical substances have bonds (and corresponding bond energies) between atoms that decompose at different temperatures. For example, a phenolic binder decomposes at a higher temperature than an alkyd binder.

8.2.2. Differential Scanning Calorimetry

If a sample of paint is scanned with increased temperature per unit time (*e.g.*, 10°C/min or 50°F/min) plot of heat capacitance versus temperature, changes in heat capacitance will occur. Major thermal transitions can be observed, such as glass transition (T_g) temperature. The T_g indicates different types of binders, such as alkyd, linseed oil, or epoxy. Although temperature transitions do not indicate chemical substances, they can differentiate between thermoplastic and thermoset (cured) binders, since the former possess a melting temperature (T_m), and the latter possess T_g. For example, a vinyl chloride binder usually melts, while an alkyd binder does not.

8.3. HIGH-PERFORMANCE LIQUID CHROMATOGRAPHY

This method resolves differences in chemical substances dissolved in a solvent. When a sample is injected into a stream of solvent and flowed through a detector at a constant rate, it generates a signal with a peak that is visible on a CRT or strip chart recorder. Where the peak is generated indicates the chemical substance, while the area under the peak indicates the concentration. With this method, substance resolution can be accomplished at moderate temperatures and substances do not have to be volatile, as they do in gas chromatography. Solvents, resins, and additives can be identified by this method, *e.g.*, analysis of a liquid paint sample for binder and solvent.

8.4. GEL PERMEATION CHROMATOGRAPHY

This method resolves different chemical substances in an unknown sample, such as binders, solvent, and additives, by molecular weight. When a sample is injected into a continuous solvent steam through a detector, its molecular weight varies inversely with retention time (time to generate signal). The area under the peak(s) on the CRT indicates the concentration. The binder has a significantly larger molecular weight than the chemical substance, whose molecular weight can be estimated by average number molecular weight (M_n) and weight average molecular weight (M_w). An application of this method involves analyzing the solvent in a liquid paint sample.

8.5. GAS CHROMATOGRAPHY

This method resolves chemical substances if the materials volatilize to form gas or vapor. A gas or vapor sample is injected into a gas stream flowing through a detector. The sample generates a signal according to the sample's chemical properties and the time of signal; the area under the peak indicates the concentration of each chemical component. For example, a sample of paint can be heated to vaporize solvent above the solution (head space), and a sample of the solvent can be taken for gas chromatography analysis. Analysis temperatures must be high so that the sample forms a vapor or gas. An application of this method involves separating components (binder, solvent, and additives), then using the molecular weight to analyze liquid paint samples.

9

Analyzing Dust and Soil for Lead

9.1. EPA LEAD STRATEGY

The EPA *Strategy for Reducing Lead Exposures*, from which a majority of the following information was taken, was publicly released in February 1991. This document, which was prepared during several years of interagency cooperation on lead abatement and lead exposure reduction, reflects the EPA's current regulatory and research program for lead. It is intended to complement similar documents produced by the Center for Disease Control (CDC) and Housing and Urban Development (HUD). The document will continue to evolve as more is learned from current research programs. Nevertheless, the strategy calls for efforts to reduce significantly (1) the incidence of blood lead levels above 10 μg/dl in children while taking into account the associated costs and benefits and (2) unacceptable lead exposure anticipated to pose a risk to children, the general public, or the environment. A discussion of this document follows.

9.2. EPA/OFFICE OF RESEARCH AND DEVELOPMENT RESEARCH PROGRAM FOR LEAD

The EPA's Office of Research and Development (ORD), working with EPA program offices and other federal agencies, developed a abroad lead research program to support the EPA lead strategy and related federal interagency activities. This multifaceted research program started in fiscal year 1990 with limited funds using congressional add-on funds appropriated to support ORD's

Multimedia Lead Research Initiative. A program was established to: (1) develop improved methods for detecting and measuring lead in paint, dust, and soil; (2) evaluate contributions to total human lead exposure from lead in various media; (3) conduct limited evaluation of existing technologies for reducing lead in paint, soil, and water; (4) evaluate bioavailability of lead from paint, dust, soil and mobilization of lead from bone to improve EPA lead uptake/biokinetic models; (5) evaluate lead health effects via support of ongoing prospective human studies of the impact of lead exposure during pregnancy or early childhood on fetal and pediatric physical/neuropsychological development; and (6) conduct various symposia or workshops on lead bioavailability factors, health effects in human adults, *etc.*), and foster other technical transfer assistance to regions/states, other federal agencies (HUD, Organization of Safety and Health Agency/National Institute of Safety and Health (OSHA/NIOSH), *etc.*), and internationally.

9.3. TESTING DUST

Household dust originates from external and internal sources. As much as 85% of the mass of household dust may come from external source, brought in by shoes and pets. The remaining fraction may be proportionally higher in lead concentration, especially if there is lead-based paint in the home. Consequently, the concentration of lead in dust may reflect soil lead, usually in the range of 100–2000 ppm in the absence of externally lead-based paint, or household dust may be overwhelmed by the influence of paint dust, which typically gives values as high as 15,000–20,000 ppm. Another factor of concern is dust loading, or amount of dust per unit area. There is some indication that total dust loading influences lead exposure more than the lead concentration. This means that dust must be sampled in a way that shows both the microgram of lead/gram of dust and the gram of dust/square meter of surface.

9.3.1. Analytic Methods

There are three commonly used analytic methods for dust analysis: XRF, flame AAS, and ICP. Preliminary results show that these methods are comparable, although not fully interchangeable. All three methods have acceptable detection limits and sensitivity, and they are linear within the normal range of environmental samples, although dilutions are usually required for the flame AAS method. Sample preparation and extraction procedures have not been standardized for any of the analytic methods.

An intercalibration study involving five labs found that each lab was able to achieve linearity with its own choice of one or two of these three instruments. Differences between methods were larger than differences between labs for each method. Five samples were used ranging from 200–3000 $\mu g/g$. An additional 15 soil samples were analyzed, and these are discussed in the following paragraphs. Although adequate instrumentation seems to be available for the laboratory analysis of lead and dust, sample preparation and extraction procedures still have problems to be resolved.

Sample preparation. Dust samples are usually small, frequently less than 200 mg. These must be sieved to remove unwanted debris and select the desired particle size. Sieving not only reduces the sample size but may also subject the small particles to separation by static electricity. Homogeneous subsampling can also be a problem.

Extraction procedure. For AAS and ICP, wet chemistry is required. Methods range from a cold, dilute nitric acid extraction to a total dissolution with nitric and hydrofluoric acids. Hydrochloric and perchloric acids have also been used, but these are less common. There is no standard extraction procedure designed to simulate human exposure.

Several commercially available spot test kits give semiquantitative results for lead in environmental samples, but these have not been fully evaluated for dust. There is at least one rhodozonate-based kit that, when used properly, gives a positive red or pink color in the presence of lead in dust. It is likely that other kits can be adapted to dust analysis, but some modifications may be required.

9.3.2. Sampling Methods

The choice of sampling methods for dust is critical. The basic methods measure $\mu g/g$ (dry vacuum) or $\mu g/m^2$ (dry vacuum and most wet-wipe methods). Human exposure is expressed as micrograms/day. Children are believed to consume about 100 mg of dust per day. Therefore, micrograms of lead/grams of dust $\times$ grams of dust/day gives the appropriate units of micrograms/day. If data become available on the effective surface area contacted by children, the wet-wipe method would be more useful.

9.3.2.1. Sampling devices

The most commonly used dust samplers are vacuum pumps with suitable filters, wet wipes, and vacuum cleaner bags. The intent is to simulate the transfer of household dust from a surface (i.e., a bare floor, windowsill, rug, toy, upholstered furniture) to the child's hand or directly to the mouth.

Vacuum pumps vary according to flow rate from about 1 l/min to over 50 l/min. Depending on the aperture, the nozzle velocity may be very high or very low and therefore over or underestimate the amount of dust transferred to the hand. Samples are taken over a prescribed ares so that both $\mu g/g$ and $\mu g/m^2$ can be determined. The sampling pattern for direction and number of passes varies with each study. For larger units, about 1 min/m^2 seems to be the accepted norm.

Wet wipes are usually purchased commercially in large lots, and there may be differences between batches for the control sample. Because the sample is often less than 10 times the blank, this can be a critical problem. Some researchers have developed techniques to produce their own wipes using filter paper and a cleaning solution. In some ways, the wet wipe accurately simulates the transfer of dust from the surface to the hand, but it does not work well on rugs or upholstered furniture, and it does not measure the mass of the dust (*i.e.*, it cannot determine $\mu \cdot g$).

Vacuum cleaner bags provide many researchers with dust samples. In general, this method has proved ineffective, largely because of the wide range in the efficiency of the typical vacuum cleaner bag. Furthermore, this method does not measure $\mu g/m^2$ and thus fails to give a measure of dust loading.

9.3.2.2. Sample scheme

The number and types of surfaces must be determined carefully to approximate effective exposure from household dust. *Hard surfaces* are the easiest to sample by either vacuum of wet-wipe methods, and the tendency is probably to oversample this type of surface because of its convenience. *Rugs and upholstered furniture* cannot be sampled with wet wipes, and a vacuum system may oversample by collecting dust not readily available to a child's hand. *Play areas* are the most logical choice for targeted sampling if this method is selected. Toddler accessibility is the first requirement, and information from the parents can also prove useful. Many researchers have found that sampling at the entry is an effective measure of *exterior dust*. Dust collections may also be made on the porch or sidewalk.

9.3.3. Standards

Such organizations as HUD, EPA, and NIST are collaborating on the preparation of standard reference materials for dust. At present, calibration standards are needed in the range of 200–20,000 $\mu g/g$ under typical conditions, however, homogeneity is difficult to achieve, and similarity to sampled dust is a critical factor.

9.3.4. Laboratory Certification

There is no certification process for laboratories testing the lead content of dust. A program has been initiated by EPA to determine the requirements for such a certification process, and it is likely that laboratory certification will be performed by an independent organization according to these guidelines. This certification will include testing for lead in dust, soil, and paint.

9.3.5. Regulatory Guidelines

There are no regulations governing the acceptable concentration of lead in household dust. In the case of paint, there are dust guidelines for clearance that seem to suggest reasonable protection from lead exposure, these are loading measurements: 200 μg/ft^2 for floors, 500 μg/ft^2 for window sills, and 800 μg/ft^2 for window wells.

9.3.5.1. Abatement

Guidelines for soil (see "Testing Soil") suggest that dust above 500 μg/g should be abated. In the context of total exposure to lead from all sources, it is possible that a level below 500 μg/g is reasonable to reduce exposure. Based on the assumption that most dust comes from soil and given that the natural concentration of lead in soil is 10–25 μg/g, it is unreasonable to expect household dust levels below this level. When there is no apparent source of dust, typical household dust concentrations are about 100 μg/g.

9.3.5.2. Clearance

Guidelines for clearance following paint abatement are probably inadequate for dust abatement. These guidelines were most likely selected because they are the lowest reasonable level attainable with paint abatement methods producing large amounts of dust. The uncertainty over dust clearance stems from the fact that there is very little information about the relationship between dust loading (μg/m^2) and lead exposure.

9.4. TESTING SOIL

Lead in soil is a mixture of lead deposited from the atmosphere, natural lead in the parent rock material from which the soil was formed, and lead from other sources, such as paint chips. Historically, the atmospheric component consisted mostly of gasoline lead in small particles (<1 μm), but more recently, with the phaseout of lead additives in gasoline, the contributors are coal-fired power plants, smelters, incinerators, as well as small amounts of lead that remain in gasoline. The residual atmospheric lead in soil may amount to as much as 2000 μg/g in areas near densely traveled highways, but typically it is less than 1000 μg/g. Natural lead in parent rock material and soil is 10–25 μg/g, soil with paint chips may reach 10,000 to 15,000 μg/g, but soil near smelters typically shows concentrations as high as 30,000 μg/g where emission and reentrainment controls were historically inadequate.

9.4.1. Analytic Methods

Analytic methods for soil are the same as those for dust in the lab: XRF, AAS, and ICP, in the field, spot test kits are being developed, as well as portable XRF instruments to survey soil-covered areas and plot the isopleths of soil concentrations.

9.4.1.1. Spot Test Kits and Portable XRF's

Spot test kits being developed for dust are also designed to indicate the presence of lead in soil. These kits will be evaluated in terms of their extraction requirements, detection limit, and the frequency and circumstances of false positives and negatives. A portable XRF instrument for soil measurements has been developed for larger sites typical of superfund clean-up operations, it is being adapted for use in an urban soil setting.

9.4.1.2. XRF, AAS, and ICP

The instrumentation and methodology for testing soils in the laboratory similar to those used for dust. Preliminary results of the intercalibration study designed to measure differences between labs and instrument methods indicate that differences between methods was greater than differences between labs for the same method.

Sample preparation for all methods consists of drying the sample to a constant weight, sieving with some crushing (usually hard crumbling), and homogenization. A hot nitric acid extraction is the most common extraction procedure for studies on human lead exposure. It is impossible to generalize on the preferred procedure because of the variability of soil types and the forms of lead in the soil. Some studies show that a cold nitric extraction over a longer period of time is comparable to hot nitric extraction. More rigorous extraction by hydrofluoric acid or perchloric acid are usually avoided because of the increased hazard and expense.

9.4.2. Sampling Methods

9.4.2.1. Sampling Devices

The choice of sampling device determines the quality of the sample. Some researchers believe that it is only important to take a sample from the surface of the soil much in the manner that a child would seize a handful of soil. However, this method does not provide information that may be helpful later in quantifying the amount of lead in relation to the depth of the soil or the horizontal distribution.

In most cases, it is important to know not only the horizontal distribution of lead at the surface, but the vertical distribution through the soil profile. This information is obtained by using a soil-coring device. This device is pushed into the soil profile to the desired depth, usually 15–25 cm. The diameter of the core is usually about 1 cm, but it may vary according to the needs of the study. If the corer meets an obstruction, it is removed, cleaned, and reinserted as close to the original location as possible.

9.4.2.2. Sampling Scheme

The sampling scheme consists of determining the number and location of specific sampling points, the number of cores to take as a composite at that point, and the depth of the sample to be taken. The usual depth of lead exposure studies is 2 cm. This assumes that lead concentrations at the soil surface is higher than at some depth and that the soil surface is in direct contact with the human environment. A core less than 2 cm is difficult to sample due to the structure and composition of the top of the soil profile. When it is important to know how far the lead extends into the soil profile, a sample is usually taken at about 15 cm.

The number of soil samples to be taken is determined by economic factors and the resolution of the study area into contoured maps of soil concentration. It is important to collect enough samples to produce a reliable estimate of exposure and provide sufficient guidance for soil abatement.

Large areas, such as parks and playgrounds, can be randomly sampled by establishing one or more straight sampling lines, identifying points at regular intervals, usually 5 meters apart, then randomly selecting the points to be sampled. Each sampling point is the center of a circle 1 meter in diameter. The circle is sampled at the center and four points on the circumference, 90° apart. These five cores are compiled into one sample. Small areas and areas targeted for specific exposure objectives (obvious play areas, pathways) are sampled in the same manner except that only one or two sample points are established.

9.4.3. Standards

Such agencies as HUD, EPA, and NIST are collaborating on the preparation of standard reference materials for soils.

Calibration standards are required in the range of 200–15,000 ppm.

9.4.4. Laboratory Certification

Due to the historically poor performance of public access laboratories, it is important to establish a certification process for laboratories as soon as possible. Protocols for this certification process have been developed, and the process of selecting a group to oversee the process is under way.

9.4.5. Regulatory Guidelines

The EPA has used interim guidance of 500–1000 ppm for abatement decisions in the Comprehensive Environmental Response Compensation and Liability Act (EPA) (superfund) and the Resource Conservational and Recovery Act (EPA) decision-making processes. There is an increasing trend to make these decisions on a site-specific basis using the EPA Uptake/Biokinetic model, which uses exposure information from all sources to develop an exposure scenario to predict blood lead distribution for the exposure population.

There are no guidelines for clearance following abatement. Replacement soil is usually less than 100 ppm, so that an effective exposure soil concentration for an area abated for all soil above 500 ppm would be 100–500 ppm.

10

General Analysis of Lead-Based Paint

10.1. ANALYSIS SCHEMES

10.1.1. General Strategy

Table 7.1 lists paint properties and methods of determinating paint composition. The analytic strategy behind investigating a paint or coating material must start with a sample of dried and/or liquid paint. A dried sample of paint must be cut from the substrate with no contamination, and a liquid sample must be taken from a well-shaken container after the pigments and ingredients have been evenly distributed.

10.1.1.1. Dried Paint Samples

The lead detection units do not require sample preparation to determine the presence of lead, but specific identification of lead pigments and other components does. A paint sample removed from a substrate should be broken, cut, or sawed so that it does not disturb the sample that will be investigated. The author recommends freezing the paint sample in liquid nitrogen, then breaking it to reveal fresh internal surfaces. It is important to remember that old dried paint is brittle. An SEM micrograph of a lead-based paint chip sample is shown in Fig. 10.1. Each layer is identified by a dividing line and noticeable differences in pigment size and color.

The presence of lead and its precise concentration in a sample can be determined by atomic spectroscopy, as discussed in Chapter 7. This method consists of digesting the weighed sample in acid, then analyzing it for lead. The

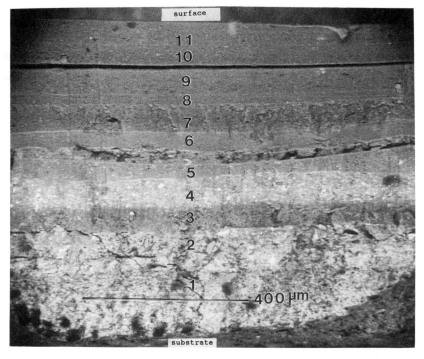

Figure 10.1. SEM micrograph of a cross-sectional view of a lead-based paint chip.

lead concentration is reported in micrograms of lead/gram of sample. If it is important to know the location of lead within the sample, then the paint sample must be cross sectioned and examined with electron microscopy.

   a. Pigments.   A transmission electron microscope and a SEM with EDXRA are excellent and rapid tools for detecting lead and other elements in individual paint layers. Although most painted surfaces consists of four or more layers, each pigment particle can be identified once the sample has been prepared to reveal the cross-sectional surfaces. Individual paint layers and pigment particles are visible in Fig. 10.1.

   b. Binders.   Infrared probe and ESCA are useful for identifying resins (binders, vehicles, *etc.*) containing pigments in a dried paint film (there are no solvents except moisture in a dried paint film). The dried paint sample can also be pulverized and digested in solvent followed by centrifugation to remove pigment particles. A film of binder particles can then be taken to generate a spectrum.

10.1.1.2. Liquid Paint Samples

Before taking a liquid paint sample, it is important first to redistribute paint ingredients, since storage produces sedimentation within the container, a paint shaker or turbine stirrer is adequate for this purpose. The analysis sample should be taken from the center of the bulk sample immediately after shaking or stirring the container. For electron microscope analysis, a liquid sample should be applied to a glass slide, frozen in liquid nitrogen, then broken to expose the fresh surfaces.

If it is necessary to identify individual components in a liquid paint, then an entire sample of liquid paint can be centrifuged to separate pigments, binders, and solvents. A successful method for accomplishing the separation follows:

1. Fill sealable centrifuge tubes (20 cm^3 or greater) with liquid paint, spin at a temperature-controlled (20 °) centrifuge at 15,000 ppm or more for 3–5 hr. Check for complete separation after 3 hr. Usually, a viscosity of about 3000–5000 cP is most successful, which means that the paint may have to be diluted with a weighted amount of solvent. Toluene has successfully been used in alkyd paints to reduce viscosity.

2. After separating individual pigments and other components, extract liquid components individually with a pipette. Weigh the sample after liquid components have been removed from pipette. Subtracting the weight of the pipette will provide an estimate of the pigment weight.

3. Remove each pigment layer with a small spatula, then weigh each layer after it has been removed to estimate the weight of each pigment type. Pigments separate by density, *i.e.*, lead pigments, which are denser than most pigments, are found on the bottom of the pipette.

4. Analyze each pigment by the previously discussed methods, such as X-ray diffraction, *etc.*

a. Binders. If the binder is mixed with a solvent, then an additional separation step is necessary. Solvents can be separated from resins by fractional distillation, providing quantitative analysis as well. Gel permeation chromatography of the combined liquid sample separates components by molecular weight, measures the molecular weight of each component, and estimates the concentration in the sample. Resins should be dried in a vacuum at 105 °C for

1–2 hr before analyzing. Dried resins can be chemically identified using infrared spectroscopy (IR).

b. Solvents. Solvents are low in molecular weight (*e.g.*, 50–100 g/mole) relative to resins (*e.g.*, 1000–10,000 g/mole). Further identification of solvent can be accomplished using gas chromatography and high-pressure liquid chromatography. Solvents can also be identified with infrared spectroscopy.

10.1.2. Reporting Results

Analysis results should be reported using standard units and citing methods of analysis. The report should include the following items.

- Calibration of instruments should be performed periodically and accompany the results of analysis.
- All information and data should be recorded in a bound notebook, each page numbered, signed, and witnessed.
- All samples should be labeled, stored, and protected for presentation of proof and possible reexamination.

10.2. DETERMINING PAINT AGE

There are basically two methods for determining paint age,

- correlating the calendar date with written records of materials
- using the isotope method.

Records of paint products must be available before a calendar date can be assigned to the manufacture, wholesale, and retail sale of paint. If these records are available, then the raw materials, including lead pigments, can be identified.

In Table 10.1, all dates (Fillippis, 1990) are reported in years before the

Table 10.1. Carbon Dating
Lead-Base Paint Components

Component	Years
Alkyd resin	7,131
Separated linseed oil fatty acid from alkyd resin	20
Lead pigment	22,550

present (A.D. 1950). By international convention, the half-life of radiocarbon is taken to be 5568 years. Standardization is with the National Bureau of Standards Oxalic Acid SRM-4990C, which is taken to be 129% modern. The reported age uncertainty is one standard deviation (68% probability). Stable carbon isotope ratios are given as both per mil difference from PDB-1 standard ration and as the calculated radiocarbon date correction in years. A radiocarbon date correction has been applied to each age to facilitate comparing different materials in nature with different carbon isotope ratios. To obtain an uncorrected date, subtract this correction factor from the reported age.

If the paint is dated without separation of components, a composite and useless date is obtained. The binder and pigments must be separated before analysis of either is feasible. The binder can be carbon dated if it is vegetable oil based, since petroleum-derived chemicals are dead or millions of years old. An alkyd resin, for example, consists of fatty acids or oils from naturally grown plants, which, when processed, rendered oil that reacted with petroleum-derived organic acids, *etc.* The fatty acid component must be separated from the binder before analysis can be accomplished by reaction of the binder with sodium hydroxide to form a sodium carboxylate on one end of the fatty acid and separate from the alkyd molecule. Having separated the vegetable-oil-derived fatty acid, it can be carbon dated (plus or minus 5 years) at the time the linseed or soya beans were harvested. This is possible, since carbon dioxide (containing carbon isotopes) in the air ceased to enter the plant. The sampling mass is a minimum of 5 g.

The same analysis conditions are true for the pigment, but dating lead carbonate is not reliable, since large figures are generated for these ages. Lead carbonate is manufactured by reacting lead with carbon dioxide in one process, so carbon dioxide may result from reaction of natural (petroleum-derived) gas in air, which would explain the very old age of the resultant carbonate anion.

10.3. DETERMINING PAINT AND PIGMENT SOURCES

10.3.1. Paint Sources

Recorded information is extremely useful for the determining the origin or supplier of paint. Written documents were generated when the paint was manufactured, including records from raw materials suppliers, and each time the product was sold and purchased. These records are in the permanent files of manufacturers, wholesalers, and retailers. A careful investigation of the manufacturer's records will reveal a batch slip that identifies each component and its source or supplier.

Table 10.2.   Examples of Manufactured Lead Pigments and Compositions

Pigment	Source	Composition/% Weight
White lead pigment (circa 1950s)	Dutch Boy Co.	1. $Pb_3(CO_3)(OH)_2$ Lead carbonate hydroxide 2. $PbCO_3$ Lead carbonate 3. $Pb_{10}(CO_3)_6(OH)_6O$ Lead oxide carbonate hydroxide 4. $Pb(BO_2)_2 \cdot H_2O$ Lead metaborate hydrate
White lead pigment (circa 1950s)	Glidden Co.	1. $Pb_3(CO_3)_2(OH)_2$/95–97% Lead carbonate hydroxide 2. $PbCO_3$/3–5% Lead carbonate
Red lead pigment (circa 1950s)	Sherwin–Williams Co.	1. Pb_3O_4 Lead (II,III) oxide 2. $Mg_3SiO_{10}(OH)_2$ Talc mica (silicate)
White lead pigment (circa 1970s)	Carter Chemical Co.	1. $3PbCO_3 \cdot 2Pb(OH)_2 \cdot H_2O$ Lead hydroxide carbonate hydrate
Red lead pigment	Carter Chemical Co.	1. Pb_3O_4 Lead (II,III) oxide

10.3.2.   Pigment Sources

Identifying the manufacturer of a pigment requires thorough investigation of the pigment components and contaminants. For example, the presence of lead metaborate was discovered by the author in lead carbonate supplied from Dutch Boy National Lead shown in Table 10.2. Since lead metaborate is not present in pigments from other suppliers, it is assumed to be a side effect of the original manufacturing process. The precise composition of lead pigment varies with lead compounds and associated contaminants, which is often very helpful. Directly comparing the composition (X-ray diffraction data) of unknown pigments with known compositions provides information on whether or not the samples are related. The degree of identification reliability depends on the quality of the known samples.

The tables in Appendix A reveal differences in each supplier's source of lead pigment. These compositions were determined by previously discussed methods.

11

Regulation and Abatement of Lead-Based Paint

11.1. LEAD-BASED REGULATIONS

Regulations for lead-based paint abatement continue to develop, and a collaborative effort of the following agencies are responsible for establishing regulations and guidelines.

- Agency for Toxic Substances and Disease Registry (ATSDR)
- Centers for Disease Control (CDC)
- Consumer Product Safety Commission (CPSC)
- Environmental Protection Agency (EPA)
- The Occupational Safety and Health Administration (OSHA)

11.1.1. Regulations for Removing Lead-Based Paint

Lead paint removal is required by different regulations, starting in 1973 when CPSC established the approved maximum lead content in paint to be 0.5% by weight (US-CPSC, 1977) in a dry film of paint. In 1978, the CPSC lowered the allowable lead level to 0.06% by weight. Housing and Urban Development guidelines require paint to be removed when lead is 1 mg/cm^2 or 0.5% by weight (HUD, 1990), 5000 parts per million, whichever is more stringent. Certain states and county authorities have adopted action levels of 1.2 or 0.7 mg/cm^2. The action level for lead in soil has been unofficially adopted by the EPA as approximately 500 parts per million (40 CFR 50.12, 1098). If wastes generated from lead-based abatement allow lead to leach at concentrations above 5 parts per million (HUD, 1990), these wastes are hazardous.

11.1.2. Worker Safety

Clearance criteria at postabatement testing (HUD, 1990) have been prepared by the states of Maryland and Massachusetts for specific interior surfaces.

- Floors: 200 $\mu g/ft^2$
- Windowsills: 500 $\mu g/ft^2$
- Window wells: 800 $\mu g/ft^2$

The OSHA Lead Standard (29 CFR 1910.1025, 1987) established two exposure limits, both applicable over an 8-hour time-weighted average, an action level of 30 $\mu g/m^3$ and a permissible exposure limit (PEL) of 50 $\mu g/m^3$. If an employee's exposure is at or above the action level, the standard requires the employer to perform exposure monitoring, medical surveillance, training, and remove workers with blood lead concentrations at or above 50 $\mu g/dl$ of blood. If exposure exceeds the PEL, the employer must do the following.

- Implement feasible work practices and engineering controls.
- Require the use of respiratory protection if these controls do not reduce worker exposures to at least the PEL.
- Develop a written compliance program that describes how the employer will reduce exposures to at least the PEL.
- Require employees to wear protective work clothing.
- Provide showers and changing rooms for workers.
- Launder all work clothes.
- Provide specially designed lunch rooms.
- Post appropriately worded warning signs in regulated areas.

Evaluating full-shift exposures to airborne lead involves using a personal air-sampling pump, calibrated to 2 liters per minute, and a Mixed Cellulose Ester (MCEF) with an average porosity of either 0.45 or 0.8 μ.

A recent report to Congress, entitled "Comprehensive and Workable Plan for the Abatement of Lead-Based Paint in Privately Owned Housing" (HUD, 1990) discusses the general operation of abatement.

11.2. ABATEMENT METHODS

The current methods for lead-based paint abatement include

- chemical paint stripping
- abrasive media cleaning

- power tool cleaning
- encapsulation
- heat gun cleaning
- enclosure
- off-site stripping, treatment, *etc.*

All of these methods have certain factors in common, such as the cost of detecting and monitoring lead substrates in and around the structure, monitoring air in the work place, and waste disposal except for encapsulation. Some of the methods and factors involved in lead-based paint removal operations are discussed in the following sections and the cost of various methods are compared.

11.2.1. Chemical Paint Stripping

Chemical-based removal methods involve using solvent-based, caustic chemicals (Dumond Chemicals, 1989; Diedrich Chemicals, 1988) or water-based strippers (3-M Corp., 1990). These can be applied manually or sprayed, depending on which products are used. Once these strippers have been applied, they are usually removed manually with putty knives or similar handtools. There are important considerations when using these materials: (1) Caustics and solvents, especially those containing methylene chloride, are potentially harmful to users or workers, so good ventilation is absolutely necessary. (2) The task may require multiple applications depending on the number of layers and type of paint to be removed. (3) Chemical paint strippers are reported to be inefficient for removing epoxy and two-part coating systems. (4) The only water-based stripper commercially available is safer to use, but it is reported to be less effective than the solvent and caustic types.

To prevent caustic chemicals from drying on substrates and tools, application and removal timing is critical.* Drying also adversely affects the removal process by increasing the possibility of creating lead dust. When using caustic chemicals, it is necessary to neutralize the substrate after removal to ensure the proper adherence of future coats of paint, and neutralization must be repeated more than once to remove any visible lead dust or residue. However, even after multiple washes, lead residues may still exist. All substrates abated by paint removal methods should be repainted and sealed with a high-gloss lead-free paint, which provides a smoother surface to clean.

---

*While caustic chemicals generally work best on metal substrates because their surfaces are usually smooth and nonporous, these chemicals should not be used on aluminum substrates with which they react.

Table 11.1. Factors Affecting
Chemically Stripping Lead-Based Paints

Factor	Costs ($ or man-hr/ft^2)
Materials costs	$1–2
Application labor	0.006–0.01 man-hr
Clean-up labor	0.02–0.03 man-hr
XFR monitoring	$0.03–0.06
Disposal of waste	$0.030–0.52

Neutralization may create large quantities of lead-bearing, liquid hazardous waste. In addition, organic solvents present a potential fire hazard from flammable vapors (such as toluene) as well as a danger to the human respiratory system, as in the case of methylene dichloride. Federal and state regulations on hazardous materials and carcinogens should be reviewed before using organic solids. Such information is contained in the product's materials safety data sheet.

Paint removal rates were discussed with manufacturers and contractors to provide the cost information in Table 11.1., and examples of chemical stripping materials are listed in Table 11.2. This type of abatement does not require skilled labor.

11.2.2. Abrasive Media Cleaning

Mechanical-based paint removal methods involve machinery and/or manual labor. Two acceptable methods on exterior surfaces are contained water blasting (Hydrosander, 1990) and abrasive blasting with a vacuum arrangement (LTC, 1990; Vacublast, 1990; Zero, 1985). Table 11.3 lists examples of vacuum abrasive media-cleaning units. Water blasting requires a containment system, which is very difficult to maintain. In addition, clean-up costs are significant, and water damage to interior structures is a possibility.

Table 11.2. Examples of Chemical Strippers
for Lead-Based Paint

Product	Manufacturer
404 RIP-STRIP®	Diedrich Chemicals
Saftest Stripper and Varnish Remover®	3M Corp.
Peel away Tile and Mastic Remover®	Dumond Chemicals

Table 11.3. Examples of Vacuum-
Abrasive Cleaning Units

Unit	Manufacturer
LTC Unit model 1030	LTC International, Inc. 101-G Executive Dr. Sterling, VA 22170 (703) 742-7860
Vacu-Blast model SB03	Vacu-Blast Corp. 125 Market St. Kenilworth, NJ 07033 (908) 245-6363
Zero C150	Zero Mfg. Co. Washington, MO (314) 239-6721

Abrasive blasting can only be used in conjunction with a vacuum arrangement. Care must be taken so that the configuration (diameter of the nozzle and the working distance from the substrate) of the blasting nozzle matches the force of the vacuum so that the system is effective in containing debris. A major disadvantage of vacuum blasting a surface is the slow rate of paint removal. The nozzle is usually placed at less than 5 in. from the surface, so that a small area less than 5 in. is affected. Consequently, vacuum blast cleaning takes much longer than conventional abrasive blasting.

Substrates that can be vacuum blasted are wood, metal (steel, aluminum, and galvanized steel), and concrete; even most irregular surface contours can be cleaned. This technology requires skilled labor.

11.2.2.1. Controlled Investigation of Vacuum-Abrasive Media Cleaning

A controlled investigation (Gooch, 1991) was conducted at Georgia Tech Research Institute to evaluate different vacuum-abrasive media cleaning on wood, aluminum, galvanized steel, and steel substrates.

Samples of actual lead-based paint on substrates were obtained from the Macon Housing Authority in Macon, Georgia (Gooch, 1991). Old paint usually consists of pigmented oil-based alkyd paint that is usually brittle, multilayered, and often cracking and chipping. Old base layers of alkyd paint have often been painted over with other types of paint, such as water-based acrylic latex. Painted wood samples consisted of shelves, windowsills, soffits, and doors; metal samples consisted of vertical and horizontal metal supports and hand rails. The majority of the coatings were alkyd- and acrylic-type binders.

Coating thicknesses were measured with a Tooke Inspection Gauge. All of the samples had multilayers of paint (not all lead pigmented), as revealed by scanning electron microscopy, which identified individual layers of paint, and energy dispersive X-ray analysis (EDXRA), which determined the presence of lead in individual layers (see Table 11.4).

Removal rates for each vacuum-abrasive cleaning unit were different due to differences in size. The smallest vacuum-abrasive cleaning unit was selected from each manufacturer in anticipation that a smaller unit would be more maneuverable within a house; however, hoses on the vacuum-abrasive cleaning units can be extended to reach other rooms, which makes it possible to use larger cleaning units. All of the vacuum-abrasive cleaning equipment in this study are similar with regard to operation and productivity; they are mobile and can be used in residential structures. The units are listed as A, B, and C in Table 11.5.

These cleaning units were evaluated for their individual qualities and compared their ability to remove lead-based paint. The maximum brush (vacuum cup) diameter indicates the relative cleaning rate for each unit, since the brush is in contact with the paint and the abrasive material flows within the brush. The actual cleaning rate is a function of the operator's technique and the effectiveness of the abrasive material.

11.2.2.2. Results of Vacuum-Abrasive Media Cleaning

Painted aluminum, galvanized steel, steel, pine wood, and particle board substrates were used for testing purposes, and they were all well cleaned, although at different rates. The paint is the critical factor for cleaning since the substrate is not purposely abraded or removed. All metal surfaces cleaned at the same rate with the same equipment because the paint was similar, but the heat diffusion property of metal surfaces reduced the heat gun's cleaning rate.

Table 11.4. Paint Materials and Characterization

Sample	Lead Content by XRF mg/cm^2	Total Layers/ Lead Layers	Total Thickness (mil)
Interior particle board shelving	4.2–4.4	3/1	3–6
Exterior pine wood	3.8–4.0	4/1,2	5–7
Pine wood window frame	4.0	4/1,2	5–7
Exterior steel columns	2.1–2.9	8/1,2,4	8–10
Aluminum soffet vents	2.5	2/1	3–5
Galvanized steel drip cap	2.2	2/1	3–5

Table 11.5. Vacuum-Abrasive Cleaning Equipment

Unit (gauge)	Air Pressure (psig)	Brush Dia. (in.)	Weight (lb.)	Cost ($)
A	100	3	75	4000
B	100	2	50	3235
C	100	3	40	3936

Table 11.6 describes the abrasive media and Table 11.7 gives the results of vacuum-abrasive media cleaning. The test patch size in each case was 3 × 12 in., and at least 10 patches were tested by each method.

Abrasive media were selected on the basis of safety, efficiency, and removal cost for lead-based paint. All media were selected because of their successful histories in abrasive cleaning operations. Silica sand was initially used, but further study revealed a potential health problem to workers, referred to as silicosis; material safety data sheets are contained in Appendix A. This condition arises when silica inhaled by workers results in lung disorders. Because even under vacuum-abrasive cleaning conditions extremely fine silica particulate can enter the working area, expensive fresh-air masks and protective clothing must be worn by workers. A material comparable to silica sand in efficiency, although 67% more expensive, is staurolite sand (Starblast), which does not cause health problems was used. Plastic media (Solidstrip) was selected because it is a mild abrasive with a low density, which is advantageous for cleaning wood. Steel shot was selected because its magnetic properties make it reclaimable and it is a good abrasive cleaning material, although very dense. Coal slag (Black Diamond) was selected for its good abrasive properties and medium density. Bicarbonate of soda (Armex) was selected because it is a

Table 11.6. Blasting/Cleaning Media and Costs, July 1991

Brand Name	Mesh	Identification	Manufacturer	Surface Profile[a] (mil)	Cost ($/100 lb)
Black Diamond	20/30	Coal slag	Foster–Dixiana Co.	0.5–1	5.25
Starblast	70/100	Staurolite sand	Dupont, Inc.	0.5–1	5.25
Steel Shot	170	Steel balls	Wheel Abrator Co.	1.5–2	29
Armex	170	Sodium bicarbonate	Arm and Hammer Co.	0.3–0.5	53
Solidstrip	12/18	Plastic media	Dupont, Inc.	0.5–1	205
BX40	30/40	Silica sand	Foster–Dixiana Co.	0.5–1	3.15

Note: [a] This is the surface profile measured on metal and wood after cleaning.

Table 11.7. Results of Vacuum-Abrasive Cleaning Methods on Lead-Based Paint

Abrasive Media	Consumption of Media (lb/ft²)	Substrate Type	Rate of Removal ft²/hr	Lead Air (mg/m³)	Substrate Before/After (mg/cm²)
Coal slag	0.35	Aluminum	2	<0.01	2.5/<0.1
	0.35	Galvanized	2	<0.01	2.2/<0.1
	0.35	Steel	2	<0.01	2.1/<0.1
	0.35	Exterior wood	2	<0.01	3.8/<0.1
	0.35	Particle board	2	<0.01	4.4/<0.1
Starblast	0.35	Aluminum	2	<0.01	2.5/<0.1
	0.35	Galvanized	2	<0.01	2.2/<0.1
	0.35	Steel	2	<0.01	2.1/<0.1
	0.35	Exterior wood	2	<0.01	3.8/<0.1
	0.35	Particle board	2	<0.01	4.4/<0.1
Steel shot	0.35	Aluminum	0.7	<0.01	2.5/<0.1
	0.35	Galvanized	0.7	<0.01	2.2/<0.1
	0.35	Steel	0.7	<0.01	2.1/<0.1
	0.35	Exterior wood	0.7	<0.01	3.8/<0.1
	0.35	Particle board	0.7	<0.01	4.4/<0.1
Armex + 5% Starblast	0.35	Aluminum	0.1	<0.01	2.5/<0.1
	0.35	Galvanized	0.1	<0.01	2.2/<0.1
	0.35	Steel	0.1	<0.01	2.1/<0.1
	0.35	Exterior wood	0.1	<0.01	3.8/<0.1
	0.35	Particle board	0.1	<0.01	4.4/<0.1
Solidstrip	0.35	Aluminum	0.6	<0.01	2.5/<0.1
	0.35	Galvanized	0.6	<0.01	2.2/<0.1
	0.35	Steel	0.6	<0.01	2.1/<0.1
	0.35	Exterior wood	0.6	<0.01	3.8/<0.1
	0.35	Particle board	0.6	<0.01	4.4/<0.1
BX40 silica sand	0.35	Aluminum	2	<0.01	2.5/<0.1
	0.35	Galvanized	2	<0.01	2.2/<0.1
	0.35	Steel	2	<0.01	2.1/<0.1
	0.35	Exterior wood	2	<0.01	3.8/<0.1
	0.35	Particle board	2	<0.01	4.4/<0.1

mild abrasive material and soluble in aqueous solutions (and therefore disposable in municipal sewers), leaving a smaller mass of hazardous lead waste. Aluminum oxide was considered for this study, but it is too dense (which increases disposal costs) and expensive for these applications.

The mesh size of each abrasive material was selected from initial experimentation with different mesh sizes to obtain the best cleaning quality at minimum destruction of the substrate surface. In the visual study comparing abrasive materials, each material was used for a half-hour to clean a surface, which was the period required for the best materials to remove 100% of the paint from metal or wood surfaces, and photographs were made.

Starblast, Black Diamond, and BX40 silica sand media performed well for these vacuum-abrasive cleaning units. These materials flow well in the cleaning units and remain abrasive until too contaminated with paint chips to be useful. Test patches on a steel column (3-in. diameter) were cleaned using Starblast and Black Diamond abrasive materials, respectively. No paint or lead was present on these surfaces, and the surface possessed a 0.5–1 mil surface profile, which required light sanding before painting. Another section of the same painted steel column was cleaned using each of the other abrasive media.

Steel shot is too dense for practical purposes to use in vacuum-abrasive cleaning equipment except in larger industrial units. The surface profile of the cleaned surface was a rough-textured 1.5–2 mils, which requires significant sanding before painting.

Armex is too soft even when mixed with 5% Starblast and requires much more air pressure than these cleaning units can accommodate. An incompletely cleaned surface resulted from using Armex. The surface profile of the surfaces were 0.5 mil. Possibly the acid-neutralizing (basic) properties of sodium bicarbonate could effectively neutralize acidic chemical paint strippers when mixed with a medium (such as Starblast) and serve as a post-cleaning/ neutralizing step.

Solidstrip is a marginal abrasive material about four times as expensive as Starblast and Black Diamond. A surface prepared with Solidstrip left residual paint. The surface profile of the cleaned surfaces were 0.5–1 mil.

Silica Sand BX40 is an efficient and cost-effective abrasive medium except for the potential health problems. A vacuum-abrasive cleaned area on steel yielded results similar to those obtained with Starblast and Black Diamond. The surface profiles of metal and wood substrates after cleaning ranged from 0.5–1 mil, which requires light sanding before painting.

In the next study, observations are based on experimental and practical experience with small vacuum-abrasive units considered most useful for residential structures. Each of the abrasive materials was evaluated for removal rate on a range of substrates with each vacuum-abrasive cleaning unit, and results are listed in Table 11.7. Each abrasive material was used until the surface was 100% clean, or as close as possible, to evaluate removal rates of each abrasive material. In order of decreasing rates of removal, the abrasive materials are (1) Starblast, Black Diamond, BX40 silica sands; (2) Steel Shot; (3) Solidstrip; and (4) Armex.

The consumption of abrasive media was evaluated, but it was soon learned that thousands of square feet of painted surface must be cleaned before determining an accurate mean value for each abrasive material. The averaged lb/ft^2 figure is listed for each material because no abrasive material was worn out during the course of this study, and estimates are based on experimental results. Starblast, Black Diamond, and BX40 silica sand are expected to become too contaminated with paint chips to be effective before the abrasive material is actually worn out or rendered structurally ineffective. Processing the mixture of abrasive material and paint chips to reclaim the abrasive material would extend use of the abrasive material and reduce the mass of disposable hazardous waste. However, the reclaiming process and equipment require an additional cost that would have to be compared to disposing of the total mass of the abrasive material and paint chip mixture. The total size of a lead paint removal contract determines whether or not to use reclaiming equipment.

Particle board, pine wood, aluminum, galvanized steel, and steel were efficiently cleaned using vacuum-abrasive cleaning, but the zinc coating on galvanized steel was only partially removed—it had a spotty appearance. Therefore, it is important to retain the thin (about 1-mil) zinc coating on galvanized steel, then another method of paint removal is recommended, such as a heat gun or chemical-stripping to remove the majority of the paint, followed by a very light surface cleaning with a vacuum-abrasive medium. Cleaned aluminum surfaces were bright in appearance and smooth; their average surface profile was 1 mil.

The vacuum-abrasive cleaning units are mobile and capable of being used inside residential housing in this study, but each manufacturer also produces larger units for industrial applications. Table 11.8 lists estimated abatement costs for this study.

The film thickness of the paint on each substrate was not so important as its adhesion to the substrate, which was greater for aluminum, galvanized steel, and steel than pine and particle board.

Regardless of material, any flat or round surface can be cleaned with the proper brush accessories, which produce a surface requiring only light sanding before painting. Wood or metal surfaces with large pits or surface deformities create a cleaning problem due to the inaccessibility of the irregular surfaces.

Table 11.8.   Cost Estimates of Vacuum-Abrasive
Cleaning of Lead-Based Paint

Equipment	Substrate	Media ($/ft²)	Labor ($9.25/hr.)	Disposal ($/ft²)	Total ($/ft²)
Units	Aluminum	0.0183	4.62	0.005	4.64
A, B, and C	Galvanized	0.0183	4.62	0.005	4.64
	Steel	0.0183	4.62	0.005	4.64
	Exterior Wood	0.0183	4.62	0.005	4.64
	Particle Board	0.0183	4.62	0.005	4.64

The general safety of each vacuum-abrasive cleaning unit is reflected in the lead-monitoring data in Table 11.7. Lead levels near the operator were below the detection limit which means that the working environment is very safe for operators and other workers. With the exception of silica sand, none of the discussed abrasive media pose a potential health problem. Nevertheless, as a precaution, the operator of vacuum-abrasive cleaning equipment should use a dust-filtering (below 1 μ) mask that fits over the nose and mouth.

Since the machines used were pneumatic, they did not require electrical power near the work. Air pressure was supplied by compressors outside the shop and can be similarly supplied in residential housing. The vacuum abrasive machines can be operated inside or outside a house with complete safety.

The maximum distance from the vacuum cup and the substrate surface is critical, since it was demonstrated that abrasive media mixed with lead-based paint could leak into the work area. More specifically, when the vacuum cup was suddenly withdrawn more than 2 in. from the substrate surface, particulate fell on the floor. This potential problem could be eliminated by installing a vacuum pressure switch near the cup so that when the cup is withdrawn more than 2 in. from the surface, air pressure supplying the flow of abrasive media could be shut off.

11.2.3. Power Tool Cleaning

A vacuum applied to such power tools as needle guns (Kurt, 1991) and others (Bloemke, 1991) eliminates most of the dust from the working environment when cleaning steel structures; however, power tools (needle guns, brushes, sanders, and others) have difficulty with nonplanar surfaces, such as corners and edges, and wood is particularly vulnerable to harsh mechanical treatment, so careful consideration should be given before using power tools.

A protective mask (fresh air or equivalent) is recommended for workers removing lead-based paint with power tools. Cleaning rates of 14–18 ft.^{2}/hr were reported by Bloemke at a cost of \$3.05/ft.^{2} At the time of writing, reliable figures for lead-based paint abatement using hand-held power tools in residential structures is not available, but they are being generated.

11.2.4. Encapsulation

11.2.4.1. Adhesive-Applied Fiberglass Mesh Encapsulants

Appropriate surfaces for encapsulation include interior walls and ceilings. Unsuitable surfaces are those subject to any kind of abrasion, such as floors, window jambs or sashes, or door jambs or edges. To apply fiberglass mats, which usually come in rolls approximately 50 in. wide, a layer of mastic or adhesive materials is first applied with a roller or brush. Then fiberglass mat is embedded in the wet mastic material, smoothed to remove bubbles, and an additional coat of mastic or adhesive material is applied over the surface, generally using the fiberglass as a mil gauge. When all gaps and crevices in the fiberglass are filled, the thickness of the mastic material is usually sufficient. It is imperative to follow the manufacturer's written instructions. Table 11.9 lists examples of encapsulant material.

No attempt should be made to apply fiberglass products (Flexi-Wall Systems, 1990) over large gaps or holes, since the product could be easily punctured, reexposing lead coatings underneath. Similarly, fiberglass products should not be applied over badly deteriorated surfaces, such as exposed brown-coat plaster unless repairs are first undertaken. The mastic coating will not adhere to the brown coat.

A test patch should be made before applying fiberglass products, and the test patch should be thoroughly cured (resins must be completely dried or reacted) according to the manufacturer's instructions. If attempts to remove the test patch yield only very small pieces, then acceptable cohesion and adhesion have been achieved.

It is not advisable to wrap fiberglass around outside corners; instead the material should be cut to fit. An outside corner bead of wood or metal is necessary to prevent delamination of the material due to abrasion. When metal is used, it must be feathered out with a joint compound.

Fiberglass materials generally require at least one or two additional coats of paint for an acceptable aesthetic finish. This technology requires skilled labor.

11.2.4.2. Paint-Supplied Encapsulants

Flexible varieties of polymer paints (or coatings) (Hampton Paint Co., 1990) are often latex based. Because of the water base inherent in latex coatings

Table 11.9. Examples of Encapsulating Materials

Encapsulant Material	Manufacturer
Composites	
Encapolastic	Audax Protective Coatings
Encap mesh	Encap Systems
Coatings	
Primer/topcoat	Hampton Paint
Two-coat system	Cente–Plast of American

systems, these products may be less toxic to both the worker and future residents; however, it is wise to submit an ingredients list to a chemist or environmental authority for a toxicity evaluation.

Surface preparation and proper application are critical with polymer materials, as are temperature and humidity of the substrate and atmosphere. Surfaces must be clean, free of chipping or peeling paint, and deglossed using a chemical deglosser or light abrasive material. Machine sanding should not be used for deglossing unless the user is certain that the substrate (coated or otherwise) does not contain lead. Materials must be applied to the proper thickness as per the manufacturer's instructions. Typically these materials are applied at wet film thicknesses (WFT) or 13–14 mils and dry to a film thickness (DFT) of 10 mils or greater. A mil gauge is a useful tool for determining the DFT of the wet coating. Temperature of the substrate and atmosphere must be a minimum 40 °F (4.4 °C).

Polymer paints must be cohesive and adhesive to the substrate, comparable to a sheet of rubber cemented over a solid substrate. Pull tests should result in tearing small bits of the paint from the substrate, thereby indicating that adhesion to the substrate is greater than cohesion within the paint film. Material cohesion versus surface adhesion must be tempered with an understanding of the particular material with which the user is working. Cohesion must be adequate so that the polymer paint will not chip and flake like a low-performance decorative-type paint. Some materials that seem to have considerable elasticity or cohesion tend to lose these properties over a period of time when subjected to freeze–thaw conditions.

There are polymer paints or similar products marketed as lead paint encapsulants that do not characteristically have a high degree of elasticity or flexibility. Because of the brittle nature of these products, they tend to chip and flake with substrate deterioration. Only products whose effectiveness has been demonstrated by a neutral and independent authority, other than the manufacturer, should be used as encapsulants. This technology requires skilled labor. Table 11.10 lists the factors influencing encapsulation costs.

11.2.5. Heat Gun Cleaning

Heat-based removal methods (Black and Decker, 1990; MHT Products, 1990; Masters, 1990) involve applying a localized heat source, such as a heat gun, to a specific surface.* This causes the paint to separate from the substrate so that it can be mechanically removed with putty knives, *etc.*; Table 11.11 lists examples of heat guns.

---

*Wooden substrates should be sanded or otherwise prepared prior to refinishing, since the wood surface could be charred during paint removal.

Table 11.10. Cost Factors
of Encapsulation

Factor	Costs ($/ft^2)
Materials	1.20
Labor	
Interior	2.50–3.50
Exterior	3–4
Average	3.70–5.20

High levels of air-borne lead can be produced by heat guns, and at the temperatures expected to occur during paint removal, some lead and organic material are likely to be aerosolized and volatilized (especially above about 480 °F (248 °C)). More specifically, a typical lead-based alkyd coating on a substrate begins to decompose above 400 °F (204 °C) by first breaking the chemical bonds of the alkyd and destroying the binder (which produces fumes) between the lead pigment particles (0.01–5 μ in diameter).

To counter the fine lead pigment particles that become airborne, a respirator with an organic vapor and particulate cartridge is strongly suggested for use by the operator. The cartridges should be capable of removing lead particulate and associated fumes, and removal efficiencies should be documented with actual test data, a fresh air mask could also be used.

In summary, heat-based methods are most appropriate for localized areas or touch-up removal of some lead-based paints excluding epoxy paints. Wide use of this method is not recommended due to the significant amount of material that becomes air-borne, creating a health hazard.

The removal rates and associated costs of this method will not be explored since it is not a method which will actually be used. This technology requires unskilled labor.

Table 11.11. Examples of Heat Guns

Unit	Manufacturer
Black and Decker model 6750	Black and Decker
Milwaukee model 750	MHT Products
Master HG501A	Master

11.2.5.1. Cleaning Lead-Based Paint by Heat Gun

All metal substrates acted as a "heat sink" for the heat guns, which reduced the heating rate and temperature of the paint. Ten test patches of 3 × 12 in. were made in each case of the following for three heat guns identified as units A, B, and C.

A "time to reach blister temperature" for paint removal was used to determine the relative speed of paint removal. In each case, the paint consisted of an alkyd resin binder on painted metal and wood surfaces. A calibrated thermocouple attached to a unit with a digital temperature scale was placed directly on the surface, and the temperature was recorded when the paint blistered; Table 11.12 lists the results. This test showed that a metal substrate requires more cleaning time (+5–72 sec) and a higher temperature (+77–82 °F (+22–28 °C)) than less thermally conductive substrate, such as wood.

11.2.5.2. Results of Heat Gun Cleaning

All test patches showed signs of overheating the wood, as shown by darkened areas. Particle board proved to be the most heat-sensitive substrate. Charred areas were obvious in all three treated test patches, but unit A produced the greatest amount of charring.

Referring to the data in Table 11.13, heat gun cleaning is faster than vacuum-abrasive cleaning for wood but slower for metal; however, lead is not sufficiently removed from substrates by heat guns. Therefore, heat guns are not viable options alone for removing lead-based paint. The heat gun method employed consisted of blistering the paint while scrapping the surface. The scrapping tool was attached to the nozzle of the gun. A thin residual film of paint was visible but could not be removed without severely charring the particle board surface. No visible smoke was observed during this process, but air monitoring near the operator revealed an average 0.04 mg/m³ of lead in the air. The averaged level of lead in the air was the same for all substrates, which indicated that lead entered the air during heat gun treatment. If many

Table 11.12. Substrates and Blister Temperature/Time

Unit	Aluminum, Galvanized Steel, and Steel Substrates	Particle Board and Pine Wood Substrates
A	414°F (212°C)/25 sec	392°F (200°C)/20 sec
B	464°F (240°C)/100 sec	424°F (218°C)/28 sec
C	446°F (230°C)/15 sec	410°F (210°C)/10 sec

Table 11.13. Results of Heat Gun Cleaning on Lead-Based Paint

Electrical Consumption[a] (w/ft²)	Substrate Type	Rate of Removal (ft²/hr)	Air (mg/m³)	Lead Substrate Before/After (mg/cm²)
0.28	Exterior wood	3.57	0.04	3.8/1.7
0.28	Particle board	4	0.04	4.4/2.8
1.12	Aluminum	1	0.04	2.5/1.7
1.12	Galvanized	1	0.04	2.2/1.7
1.12	Steel	1	0.04	2.1/1.1

heat gun units were operated together in a residential structure, then the cumulative effect could be unsafe for operators. In any case, caution is urged for large-scale use of heat guns.

Monitoring air in the work area was acceptable as long as working temperatures (Table 11.12) were not significantly surpassed. Higher temperatures generated smoke and unacceptable lead particulate levels in the atmosphere. The following items are important to remember when using heat guns.

- Temperatures must be monitored and no smoke can be generated; proper temperature can be maintained by selecting the correct heat gun model and electrical power.

Table 11.14. Cost Estimates of Heat Gun Cleaning Lead-Based Paint

Equipment	Substrate	Labor Cost ($/ft² at $9.25/hr)	Disposal Cost ($/ft²)	Total Cost ($/ft²)
Black and Decker model 6750	Aluminum	9.80	0.001	9.80
	Galvanized	9.80	0.001	9.80
	Steel	9.80	0.001	9.80
	Exterior wood	2.59	0.001	2.59
	Particle board	2.31	0.001	2.31
Milwaukee model 750	Aluminum	9.25	0.001	9.25
	Galvanized	9.25	0.001	9.25
	Steel	9.25	0.001	9.25
	Exterior wood	2.59	0.001	2.59
	Particle board	2.31	0.001	2.31
Master model HG501A	Aluminum	6.10	0.001	6.10
	Galvanized	6.10	0.001	6.10
	Steel	6.10	0.001	6.10
	Exterior wood	2.31	0.001	2.31
	Particle board	1.57	0.001	1.57

- Metal substrates require much longer periods of time for paint removal than wood substrates due to the diffusion of heat into the metal.
- Protective breathing devices that filter lead particulate and adsorb products of decomposition should be worn by all operators.

11.2.5.3. General Safety

Referring to data in Table 11.13, the substrates were cleaned with extreme care taken not to overheat the paint. From experiments, the heat gun method of cleaning substrates depends on the operator's skill, and some air-borne particulate is unavoidably generated. A dust-filtering mask is mandatory for all operators and personnel in the work area; the mask must be tested to ensure protection. The heat gun method is generally evaluated as marginal, and much "caution" should be employed during its use. Table 11.14 lists the cost of heat gun stripping.

11.2.6. Combined Cleaning Methods

11.2.6.1. Composite Heat Gun–Vacuum-Abrasive Cleaning

A combination of heat gun paint removal followed by light vacuum-abrasive cleaning is faster than the vacuum-abrasive method alone, and removal rates with the composite methods are significantly higher for both wood and metal substrates. Light abrasive cleaning of heat-gun-treated particle board yielded a paint-free surface without serious roughness, which required only light sanding to be ready for painting. Slightly charred areas were visible from heat gun cleaning.

A lead-based painted (alkyd paint) steel substrate was heat gun and then vacuum-abrasive (LTC Model 1030) cleaned using Starblast abrasive material. The paint was blistered but not completely removed to a lesser extent on steel than on the previously cleaned particle board. Vacuum-abrasive cleaning with Starblast abrasive material after heat gun application removed the paint and prepared the surface sufficiently for repainting.

11.2.6.2. Composite Paint Removal Methods and Their Costs

The combination of heat gun and vacuum-abrasive cleaning provides the most labor- and cost-effective method of lead paint removal. The heat gun method removes paint faster from wood substrates, but it does not provide a lead-free substrate as vacuum-abrasive cleaning does. The rate of removal for different substrates is shown in Table 11.15. Table 11.16 shows that vacuum-abrasive cleaning is most cost effective for cleaning aluminum, galvanized

Table 11.15. Results of Composite Heat Gun–Vacuum-Abrasive
Cleaning of Lead–Based Paint

Equipment	Substrate Type	Rate of Removal Heat Gun/Vacuum Cleaning (hr/ft²)	Total Removal Rate (ft²/hr)
Heat Guns and Vacuum- Abrasive Cleaning	Pine wood and particle board	(B&D)[a] $0.28 + 0.08$ (LTC)[b] $= 0.36$	2.78
	Aluminum	(B&D)[a] $1.06 + 0.20$ (LTC)[b] $= 1.26$	0.79
	Galvanized	(B&D)[a] $1.06 + 0.20$ (LTC)[b] $= 1.26$	0.79
	Steel	(B&D)[a] $1.06 + 0.20$ (LTC)[b] $= 1.26$	0.79

Notes: [a]Black and Decker Model 6750.
       [b]LTC International Model 1030.

steel, and steel, while the composite method is more cost effective for cleaning pine wood and particle board.

A significant benefit of the composite method is a reduction of waste material by virtue of removing the bulk of the paint without having to use abrasive media for residual paint. This advantage reduces hazardous waste disposal costs by as much as 50%.

11.2.7. Summary

11.2.7.1. Comparison of Methods

Based on these tests, we conclude that chemical paint stripping is a practical and safe method of removing lead paint providing the stripper does not contain such volatile solvents as methylene dichloride or other materials such as

Table 11.16. Costs of Composite Heat Gun–
Vacuum-Abrasive Cleaning

Substrate	Media ($/ft²)	Labor ($9.25/hr.)	Disposal ($/ft²)	Total ($/ft²)
Aluminum	0.0183	14.43	0.005	14.45
Galvanized	0.0183	14.43	0.005	14.45
Steel	0.0183	14.43	0.005	14.45
Exterior wood	0.0183	3.33	0.005	3.35
Particle board	0.0183	3.33	0.005	3.35

caustic solutions that could harm the user. A water-based stripper is acceptable in areas that are not subject to damage by excessive water, since the stripper must be exposed to the paint for up to 24 hr. In such areas, caustic base strippers with nonvolatile solvents are acceptable. A neutralizer is required with these strippers due to the high pH chemicals involved; this adds a step to the process. Water-based strippers are reported to be slower than caustic- and solvent-based strippers. Chemical paint strippers have the disadvantage of not being capable of removing lead particles imbedded in wood grain or voids in metal.

Only vacuum-abrasive cleaning methods are acceptable, since dust and large amounts of water are undesirable. The cleaning rate depends on the operator and the target area under the blaster, which is about 4 in. in diameter. Pentek has reported an abrasive–vacuum method using a needle gun to chip the paint, which is then removed with a vacuum system; airborne emissions are below 30 μ/m^3. Vacuum-equipped power tools, such as rotary brushes, hammers, and grinders may be viable options, but these had not been investigated at the time of writing.

Encapsulating lead paint surfaces is an alternative to removal methods, and several products that have been used successfully are available at a range of prices.

A combination of heat gun stripping and abrasive media cleaning is a faster and more effective method than either method used alone. Heat gun stripping, although faster, does not remove 100% of the paint, while the slower abrasive media cleaning method removes 100% of the remaining paint.

The condition of the treated surfaces after abatement is important, since these surfaces must be refinished for actual use. The chemical stripping operation does not cause damage but requires neutralization. Encapsulation does not require refinishing, and the abrasive cleaning methods requires minimal touch-up sanding to smooth out the surface.

From discussions with contractors involved in lead paint abatement with city housing authorities, it appears that chemical paint strippers are most widely used. Reliable standardized costs per square foot or other measures of cost are not available, and only job-by-job information has been reported. Usually, a job is quoted by the contractor, which includes materials, labor, and other factors affecting the total cost. Table 11.17 lists the advantages and disadvantages of the methods reviewed.

11.2.7.2. Cost Comparisons

Direct cost comparisons are not feasible on a national basis, since the cost of labor varies from state to state. A cost analysis of abatement methods is

Table 11.17. Major Advantages and Disadvantages
of Lead Paint Abatement Methods

Abatement Method	Advantages	Disadvantages
Chemical paint strippers	No dust Requires unskilled labor No dust Inexpensive	Does not perform on all paints Requires waste clean-up and disposal May require protective clothing/respirator
Vacuum-abrasive cleaning	No dust No clean-up Performs on all paints	Requires waste disposal More expensive than chemical stripping Performs marginally on odd-shaped surfaces Requires skilled labor
Vacuum power tool cleaning	No dust No clean-up Performs on all paints	Requires waste disposal Performs poorly on odd-shaped surface
Encapsulation	Inexpensive No dust Performs on all paints No refinishing No clean-up No waste disposal	Leaves lead on structures Requires skilled labor
Heat gun cleaning	Fast	Produces fumes and lead particulate Leaves lead on surfaces
Vacuum-combination heat gun/abrasive cleaning	Fast and effective	Requires two methods

based on information obtained from a project involving Georgia Technical Research Institute and the Macon Housing Authority. Table 11.17 lists some preliminary results from this project for units of approximately 90 ft², a range of total cost per generic method is listed in Table 11.18. It is important to understand that encapsulation methods include refinishing so no painting nor covering is necessary, whereas the other methods require refinishing at an additional cost. Therefore, encapsulation is actually less expensive than any of the other methods; the only disadvantage is that the lead paint remains within the building. Comparative figures are shown in Table 11.10; the figure for encapsulation is for fiber-reinforced resin products.

There are differences in lead paint disposal methods; for example, abrasive cleaning produces finer particles, which are capable of leaching lead at a faster rate, so the cost of waste disposal is high. However, abrasive cleaning

Table 11.18. Costs Comparisons of Lead-Based Abatement Methods

Method	Operation Labor (man-hr/ft²)	Clean-Up Labor (man-h/ft²)	Materials/Electrical ($/ft²)	Lead Monitoring ($/ft²)	Waste Disposal ($/ft²)	Total Cost ($/ft²)
Chemical paint stripping	1	1			0.05	1–2
Vacuum–abrasive cleaning	4.62	—	—	0.1–0.50	0.05	4.64
Power tool cleaning				0.1–0.50	0.05	3.05
Encapsulation	0.05–0.07	—	1.2	0.1–0.50	—	3.70–5.20
Heat gun cleaning	1.57–9.80	—	0.28–1.68	0.1–0.50	0.05	1.57–9.25
Combination of heat gun– vacuum abrasive cleaning	2.70–14.43	—	0.0183	0.1–0.50	0.05	2.29–14.45

Note: [a]These figures were collected in July 1992 from vendors and contractors, but a reliable comparison throughout the United States, with its many different labor rates and costs, is not possible except on a local scale.

methods reduce clean-up costs, since in principle, all waste is trapped in a bag. The actual cost of lead paint abatement depends primarily on the labor and equipment involved and job site variables. Information in Table 6.18 is the best available at this time. It is hoped that specifications for abatement can be established to standardize the process and eliminate indecisiveness about methods and strategy and reduce costs. Such specifications would also produce more reliable estimates of abatement costs.

11.2.8. Conclusions

Although heat guns work rapidly, they are not efficient enough to leave a lead-free surface. Vacuum-abrasive cleaning, which is slower, does leave a lead-free surface. It is safe and efficient, which is important when working in residential housing. An observation that was verified during this study was the necessity of removing a thin layer of wood from any wooden surface to assure acceptable lead removal. Apparently, lead enters the wood cells and does not always reside on the immediate surface; therefore, an abrasion method is necessary to remove lead totally from wood.

The rate of lead-based paint removal was not significantly affected by film thicknesses within the ranges indicated, since the diameter of the abrasive particles were larger than the coating thicknesses and quickly penetrated to the substrate surfaces.

The zinc coating on galvanize steel was partially removed by abrasive cleaning, but there was no roughening effect on the surface. None of the other substrates were significantly roughened by abrasive cleaning when the correct abrasive material was used.

Labor is the most significant cost after equipment; however, labor could become the largest cost for a contract encompassing thousands of square feet of surface for abatement. Average labor cost is estimated at $9.25/hr (Macon, Georgia, June 1991). A cost-effective method of paint removal must be the most labor-efficient method.

11.2.8.1. Interior Surfaces

Chemical paint stripping is recommended where it is cost effective. Vacuum-abrasive cleaning should follow chemical paint stripping and heat gun treatment because a residual amount of lead remains on the substrate with the two latter methods. If the heat gun treatment is not carefully controlled, the paint may become too hot, causing the paint to decompose and produce fumes; as a result, lead particles enter the atmosphere. Most interior surfaces are particle board, wood, or sheet rock. A test patch should be made to determine if a heat gun (or other method) can remove the bulk of the paint. If so, the heat gun treatment should be followed by vacuum-abrasive cleaning. If the heat gun alone has no effect, then vacuum-abrasive cleaning should be used.

11.2.8.2. Exterior Surfaces

The methods and equipment discussed in "Interior Surfaces" also perform well for exterior parts of the structure. Larger vacuum-abrasive cleaning units may be used outdoors, since the size of the equipment is not limited by the working space.

12

Disposing of Lead Paint Waste

12.1. FEDERAL AND STATE REGULATIONS

In 1976, the U.S. Congress passed the RCRA. The RCRA directed the EPA to develop and implement a program to protect human health and the environment from improper hazardous waste management. The program is designed to control hazardous waste from its creation to its ultimate disposal (Babb, 1991). Most states run the RCRA program for EPA. In all cases, the state program must be as strict as the federal program; in some states, the state program is stricter than the federal program. North Carolina adopts the federal regulations by reference.

In 1984, the Hazardous and Solid Waste Amendments (HSWA) was passed, one major addition was the Small Quantity Generator regulations. A hazardous waste may exhibit one of four characteristics: ignitability, corrosivity, reactivity, or toxicity. Acetone, gasoline, and industrial alcohols have a flash point less than 140 °F (60 °C) and are therefore ignitable hazardous wastes. Corrosive hazardous wastes have a pH of ≤2 or ≥12.5 or corrode steel at a rate greater than 0.25 in. per year. Reactive hazardous waste is capable of reacting with air or water, causing an explosion or releasing poisonous fumes.

The toxicity characteristics (TC) constituents include heavy metals, such as lead, silver, and arsenic; pesticides; and other organics. Regulatory levels have been established for the 39 TC constituents. The toxicity characteristic leaching procedure (TCLP) is the laboratory analysis required to determine if waste exceeds the regulatory level and is thus a hazardous waste. If by the TCLP, waste is shown to leach 5 ppm of lead or greater, the waste is hazardous and has to be treated as such. Laboratory analysis is required before lead paint

221

debris can be used in sanitary landfill; if laboratory analysis reveals a high ppm, this waste would have to go to a permitted hazardous waste landfill for disposal. In North Carolina, leachable lead, as determined by TCLP, must be below 0.5 ppm to be used in sanitary landfill and between 0.5–5.0 ppm to be used in an industrial landfill. North Carolina also requires a form called the waste determination sheet to be completed and sent to the state Solid Waste Section with a laboratory analysis before lead paint debris can be used in sanitary or industrial landfill in the state.

All debris generated during the removal of the existing paints must be contained and secured until a hazardous waste determination is made. Applicable air quality regulations apply during sandblasting. Any waste generated may be separated into hazardous and nonhazardous; e.g., lead can be separated from the abrasive, but the lead cannot be diluted to render it nonhazardous. The generator had 90 days from the day the waste was determined to be hazardous to move it off site. The hazardous waste must be properly stored prior to shipment and labeled hazardous waste with the accumulation starting date.

The generator of hazardous waste must obtain an EPA identification number. In these situations, some states issue regular EPA identification numbers. The North Carolina Hazardous Waste Section issues 90-day provisional identification numbers for this one-time only waste generation. An application must be filled out, signed, and sent to the proper state agency. A list of state agencies are contained in Appendix C. The hazardous waste must be stored off site using a four-part uniform hazardous waste manifest. The transporter must have an EPA identification number, and the treater/disposer must have permission to manage hazardous waste. State agencies can provide a list of hazardous waste service companies operating within the state.

12.2. LAND BAN

Another major addition to the HSWA was the Land Disposal Restrictions (LDR) or Land Ban regulations. As a result, all hazardous waste is now banned from land disposal in most states until it is proven to leach less than 5 ppm.

A notice with the following information must accompany each shipment of restricted waste.

- EPA hazardous waste number
- Chemical and corresponding treatment standard
- Manifest number associated with the shipment
- Waste analysis data (when available)

Manifests, notices, and other records must be kept for 5 years.

Appendix A

X-Ray Powder Files of Lead Compounds

Table A.1. X-Ray Powder File, Lead

| | | | X-Ray Powder Data | | | | | |
hkl	d (Å)	I	hkl	d (Å)	I	hkl	d (Å)	I
111	2.855	100	600	0.8251	4			
200	2.475	50						
220	1.750	31						
311	1.493	32						
222	1.429	9						
400	1.238	2						
331	1.1359	10						
420	1.1069	7						
422	1.0105	6						
511	0.9526	5						
440	0.8752	1						
531	0.8369	9						

Source: Swanson *et al.* (1951).
Notes: Identification: lead
Unit cell composition: Pb
Crystalline structure: cubic

Table A.2. X-Ray Powder File, Lead 2H

			X-Ray Powder Data					
hkl	d (Å)	I	hkl	d (Å)	I	hkl	d (Å)	I
100	2.82	75	213	0.9195	15			
002	2.69	30						
101	2.51	100						
102	1.95	10						
110	1.633	70						
103	1.514	20						
112	1.397	35						
201	1.367	35						
202	1.254	5						
203	1.110	15						
211	1.047	20						
212	0.9927	10						

Source: Takahashi *et al.* (1969).
Notes: Identification: lead
Unit cell composition: (Pb)2H
Crystalline structure: hexagonal

Table A.3. X-Ray Powder File, Lead Acetate

			X-Ray Powder Data					
hkl	d (Å)	I	hkl	d (Å)	I	hkl	d (Å)	I
	11.9	450		4.97	2		3.81	12
	9.75	600		4.85	35		3.76	100
	6.93	4		4.82	10		3.67	8
	6.72	40		4.65	35		3.60	8
	6.16	16		4.55	10		3.55	25
	6.10	20		4.52	30		3.51	90
	5.89	12		4.41	10		3.42	40
	5.82	8		4.19	2		3.36	20
	5.73	55		4	55		3.33	30
	5.37	4		3.98	40		3.29	25
	5.26	4		3.92	16		3.25	8
	5.18	4		3.89	12		3.21	16
	3.15	30		2.487	30		2.181	8
	3.09	16		2.462	8		2.132	6
	3.06	40		2.447	16		2.114	4
	2.942	95		2.427	18		2.069	8
	2.895	25		2.397	20		2.048	14
	2.820	12		2.354	30		2.028	25
	2.728	4		2.337	20		1.999	14
	2.680	8		2.322	20		1.962	8
	2.642	4		2.295	25		1.940	12
	2.605	14		2.263	12		1.920	10
	2.584	16		2.240	25		1.898	8
	2.520	8		2.217	12		1.875	10
	1.855	8						
	1.833	4						
	1.818	4						
	1.758	10						
	1.731	4						
	1.702	8						
	1.660	8						
	1.654	8						

Source: Kwestroo *et al.* (1965).
Notes: Identification: lead(II) acetate
Unit cell composition: $C_2H_6O_4Pb$
Crystalline structure:

Table A.4. X-Ray Powder File—Lead Carbonate, $PbCO_3$
(Cerrusite)

				X-Ray Powder Data				
hkl	d (Å)	I	hkl	d (Å)	I	hkl	d (Å)	I
110	4.427	17	221	2.081	27	150	1.615	2
020	4.255	7	041	2.009	11	241	1.588	6
111	3.593	100	202	1.981	9	151	1.563	5
021	3.498	43	132	1.933	19	004	1.536	5
002	3.074	24	113	1.859	21	302	1.503	4
012	2.893	2	023	1.847	8	223	1.503	4
102	2.644	2	222	1.796	4	104	1.503	4
200	2.589	11	042	1.750	2	330	1.503	4
112	2.522	20	310	1.693	1	043	1.475	5
130	2.487	32	240	1.642	2	411	1.449	3
220	2.213	7	051	1.642	2	242	1.449	3
040	2.129	2	311	1.632	6	152	1.430	2
060	1.417	1						
143	1.417	1						
322	1.417	1						
332	1.330	5						
204	1.321	3						
313	1.306	5						
214	1.306	5						
134	1.306	5						
400	1.296	2						
243	1.282	3						

Source: Swanson *et al.* (1953).
Notes: Identification: lead carbonate
Unit cell composition: $PbCo_3$
Crystalline structure: orthorhombic

Table A.5.   X-Ray Powder File—Lead Carbonate, Pb_2CO_4

hkl	d (Å)	I	hkl	d (Å)	I	hkl	d (Å)	I
				X-Ray Powder Data				
	4.02	80		2.252	50		1.674	50
	3.94	80		2.215	50		1.636	50
	3.46	10		2.149	60		1.616	10
	3.31	10		2.124	20		1.601	10
	3.15	100		2.033	30		1.581	60
	3.11	100		1.982	50		1.563	50
	2.876	30		1.908	30		1.544	20
	2.801	50		1.885	50		1.498	10
	2.609	30		1.861	10		1.485	10
	2.519	50		1.801	40		1.472	30
	2.451	10		1.732	30		1.454	50
	2.297	30		1.709	60		1.428	10
	1.410	50		1.213	30			
	1.371	30		1.192	50			
	1.359	10		1.165	50			
	1.339	20		1.150	20			
	1.321	20		1.136	40			
	1.308	10		1.112	40			
	1.287	30		1.096	40			
	1.276	10		1.088	40			
	1.266	80		1.075	20			
	1.255	10		1.067	30			
	1.244	20		1.059	40			
	1.232	30						

Source: Peretti (1957).
Notes: Identification: lead carbonate
Unit cell composition: Pb_2CO_4
Crystalline structure:

Table A.6. X-Ray Powder File, Lead Carbonate Oxide,
$Pb_3O_2CO_3$, $PbCO_3 \cdot 2PbO$

| | | | | X-Ray Powder Data | | | | |
hkl	d (Å)	I	hkl	d (Å)	I	hkl	d (Å)	I
	7.06	6		2.493	2		1.712	6
	5.56	6		2.427	2		1.679	10
	4.54	6		2.345	4		1.639	8
	4.45	10		2.274	6		1.569	6
	3.99	6		2.188	8		1.517	4
	3.51	2		2.069	2		1.498	4
	3.36	4		2.040	10		1.432	2
	2.999	100		1.990	10		1.415	2
	2.934	25		1.844	4		1.410	<2
	2.869	25		1.812	4		1.385	<2
	2.777	16		1.782	10		1.307	2
	2.633	2		1.757	4		1.288	4
	1.273	2						
	1.242	2						
	1.222	<2						

Source: Pannetier *et al.* (1964).
Notes: Identification: lead carbonate oxide
Unit cell composition: $Pb_3O_2CO_3$
Crystalline structure:

Table A.7. X-Ray Powder File, Lead Carbonate Oxide,
Pb_3CO_5, $PbCO_3 \cdot 2PbO$

| | | | | X-Ray Powder Data | | | | |
hkl	d (Å)	I	hkl	d (Å)	I	hkl	d (Å)	I
	7.07	4		2.427	2		1.680	8
	5.55	16		2.368	4		1.640	10
	4.57	4		2.349	8		1.571	16
	4.46	10		2.280	8		1.504	6
	4.02	4		2.189	6		1.434	6
	3.60	2		2.048	16		1.418	2
	3.52	2		1.997	6		1.386	6
	3.01	55		1.847	10		1.292	10
	2.939	25		1.790	20		1.184	2
	2.875	35		1.759	16		1.142	4
	2.776	100		1.714	10		1.122	4
	2.659	2		1.696	6		1.109	6

Source: Grisafe *et al.* (1964).
Notes: Identification: lead oxide carbonate
Unit cell composition: Pb_3CO_5
Crystalline structure: orthorhombic

Table A.8.   X-Ray Powder File, Lead Carbonate Oxide,
$Pb_3C_2O_7$, $2PbCO_3 \cdot PbO$

			X-Ray Powder Data					
hkl	d (Å)	I	hkl	d (Å)	I	hkl	d (Å)	I
002	6.95	10	114	2.115	2	304	1.408	2
100	4.63	10	203	2.067	30	216	1.394	2
003	4.63	10	106	2.067	30		1.388	4
101	4.38	2	107	1.821	4	222	1.311	2
102	3.86	2	210	1.747	16	217	1.311	2
004	3.47	40	116	1.747	16	306	1.283	6
103	3.27	65	211	1.734	10	223	1.283	6
110	2.669	65	213	1.634	10	311	1.277	2
105	2.379	2	214	1.560	2	313	1.236	2
200	2.311	100	300	1.541	2		1.232	2
006	2.311	100	302	1.504	2		1.217	6
201	2.281	4	303	1.462	30	400	1.156	16
403	1.122	6						
	1.108	2						
320	1.061	6						
	1.040	4						
323	1.034	2						
	1.032	2						
325	0.991	4						
	0.969	6						
	0.962	2						

Source: Grisafe *et al.* (1964).
Notes: Identification: lead oxide carbonate
Unit cell composition: $Pb_3C_2O_7$
Crystalline structure: hexagonal

Table A.9. X-Ray Powder File, Lead Carbonate Hydroxide, $Pb_3(CO_3)_2(OH)_2$, Hydrocerussite

			X-Ray Powder Data					
hkl	d (Å)	I	hkl	d (Å)	I	hkl	d (Å)	I
003	7.80	5	10$\bar{1}$0	2.099	20	300	1.513	20
101	4.47	60	20$\bar{5}$	2.046	30	303	1.485	20
012	4.25	60	027	1.884	20	128	1.485	20
104	3.61	90	119	1.856	30	21$\bar{1}$0	1.388	10
015	3.29	90	101$\bar{3}$	1.696	40	201$\bar{4}$	1.353	10
107	2.715	20	21$\bar{4}$	1.649	20	111$\bar{5}$	1.353	10
110	2.623	100	021$\bar{0}$	1.649	20	121$\bar{1}$	1.340	10
009	2.623	100	12$\bar{5}$	1.613	30	309	1.309	30
018	2.491	30	001$\bar{5}$	1.584	20	220	1.309	30
021	2.261	10	111$\bar{2}$	1.576	10	223	1.292	30
202	2.231	50	201$\bar{1}$	1.562	10	312	1.251	30
024	2.120	30	21$\bar{7}$	1.530	30	211$\bar{3}$	1.251	30

Source: JK Olby (1953).
Notes: Identification: lead carbonate hydroxide
Unit cell composition: $Pb_3(CO_3)_2(OH)_2$
Crystalline structure: hexagonal

Table A.10. X-Ray Powder File, Lead Chromate, $PbCrO_4$

			X-Ray Powder Data					
hkl	d (Å)	I	hkl	d (Å)	I	hkl	d (Å)	I
101	5.43	8	210	3.15		300	2.320	
110	5.10	4	211	3.09		131	2.252	
011	4.96	4	012	3.03		103	2.252	
111	4.38	10	112	3.00		311	2.243	
101	4.37	4	202	2.710		310	2.214	
111	3.76	8	211	2.653		003	2.214	
020	3.72	8	112	2.597		131	2.154	
200	3.48	8	212	2.549		302	2.130	
002	3.32		221	2.510		013	2.120	
102	3.28		122	2.460		212	2.090	
120	3.28		301	2.351		203	2.080	
021	3.24		031	2.320		301	2.056	
312	2.046	4	232	1.829	2		1.614	6
213	2.002	8	223	1.814	6		1.601	4
231	2.002	8	303	1.806	4		1.574	2
032	1.988	4	140	1.796	6		1.552	4
321	1.988	4	141	1.760	2		1.545	6
132	1.978	20	400	1.740	4		1.534	2
320	1.967	14	302	1.734	2		1.515	2
023	1.900	8		1.710	4		1.482	2
231	1.866	4		1.694	10		1.469	2
040	1.857	4		1.658	6		1.452	2
322	1.847	25		1.641	6		1.444	2
132	1.847	25		1.621	4		1.434	4
	1.423	12		1.269	2			
	1.407	2		1.255	4			
	1.403	6						
	1.395	2						
	1.386	4						
	1.378	2						
	1.367	2						
	1.357	6						
	1.333	4						
	1.318	2						
	1.305	6						
	1.290	4						

Source: Dewolff (1948).
Notes: Identification: lead chromate
Unit cell composition: $PbCrO_4$
Crystalline structure: monoclinic

Table A.11. X-Ray Powder File, Lead Chromium Oxide, Pb_2CrO_5

			X-Ray Powder Data					
hkl	d (Å)	I	hkl	d (Å)	I	hkl	d (Å)	I
101	6.46	16	112	2.984	80	321	2.272	13
200	6.34	16	112	2.881	20	402	2.266	16
101	5.973	12	020	2.838	35	321	2.204	2
110	5.184	2	411	2.651	4	013	2.188	5
011	4.438	13	121	2.601	2	222	2.132	6
301	3.786	9	220	2.592	4	600	2.113	10
211	3.737	4	121	2.564	6	222	2.058	20
002	3.555	3	411	2.512	15	512	2.018	2
211	3.542	6	402	2.480	16	303	1.992	5
310	3.390	100	501	2.460	8	611	1.951	4
202	3.230	15	312	2.368	8	413	1.875	9
400	3.169	1	510	2.314	12	422	1.869	19

Source: *National Bureau Standards Monograph* (1977).
Notes: Identification: lead chromium oxide
Unit cell composition: Pb_2CrO_5
Crystalline structure: monoclinic

Table A.12. X-Ray Powder File, Lead Chromium Oxide, $PbCrO_4$

			X-Ray Powder Data					
hkl	d (Å)	I	hkl	d (Å)	I	hkl	d (Å)	I
011	4.40	40	220	2.347	40	303	1.841	70
200	4.33	45	103	2.288	20	031	1.801	40
111	3.92	30	302	2.247	35	123	1.771	35
201	3.70	30	221	2.231	40	131	1.763	35
002	3.56	50	022	2.199	30	322	1.749	35
210	3.42	70	400	2.166	15	230	1.710	55
102	3.29	70	122	2.128	60			
211	3.09	60	113	2.117	55			
112	2.841	45	410	2.020	40			
020	2.798	100	222	1.966	20			
301	2.675	35	321	1.932	20			
212	2.468	25	402	1.841	70			

Source: Conti *et al.* (1959).
Notes: Identification: lead chromium oxide
Unit cell composition: $PbCrO_4$
Crystalline structure: orthorhombic

Table A.13. X-Ray Powder File, Lead Chromium Oxide, $Pb_2(CrO_4)O$

				X-Ray Powder Data				
hkl	d (Å)	I	hkl	d (Å)	I	hkl	d (Å)	I
001	6.43	50	312	2.876	20	603	1.978	30
200	6.34	50	020	2.831	50	511	1.943	20
201	5.96	30	311	2.647	10		1.926	10
110	5.18	10	220	2.577	10	711	1.862	50
111	4.40	30	511	2.510	30	622	1.771	40
201	3.77	20	202	2.475	40	331	1.745	30
202	3.58	30	512	2.361	30	801	1.720	40
401	3.48	10	510	2.312	30	620	1.688	20
310	3.38	100	602	2.263	40	803	1.658	40
002	3.22	30	313	2.191	10	114	1.622	30
402	3.02	10	600	2.105	30	512	1.604	20
112	2.979	100	422	2.051	20	531	1.562	10
711	1.522	40	640	1.177	10			
602	1.499	40	1̲202	1.160	10			
911	1.465	30						
040	1.418	20						
732	1.368	20						
622	1.326	20						
005	1.292	20						
534	1.280	10						
1̲022	1.256	20b						
8̲02	1.230	10						
115	1.205	20						
1̲114	1.188	20b						

Source: Williams et al. (1970).
Notes: Identification: lead chromium oxide
Unit cell composition: $Pb_2(CrO_4)O$
Crystalline structure: monoclinic

Table A.14. X-Ray Powder File, Lead Hydroxide Carbonate Hydrate,
$3PbCO_c \cdot 2Pb(OH)_2 \cdot H_2O$

hkl	d (Å)	I	hkl	d (Å)	I	hkl	d (Å)	I
				X-Ray Powder Data				
100	7.87	20	212	2.89	70	217	2.27	70
102	6.63	50	108	2.89	70	220+	2.27	70
104	4.89	20	213	2.79	70	306	2.22	80
111	4.46	20	117	2.79	70	310	2.176	70
112	4.26	90	214	2.67	10	1110	2.176	70
113	3.97	80	300	2.61	90	224	2.132	20
202	3.72	10	302	2.55	10	314	2.058	70
106	3.66	90	118+	2.55	10	219	2.027	50
204	3.34	100	0010	2.48	90	1111	2.027	50
116	3.05	80	304	2.41	50	226	1.987	20
210	2.95	80	216	2.41	50	316	1.936	50
211	2.95	80	119	2.36	10	2110	1.905	50
404	1.870	20	326	1.653	70			
317	1.870	50	1114	1.653	70			
228	1.828	20	2014	1.617	70			
2012	1.828	20	327	1.617	70			
3010	1.801	90	2211	1.602	10			
321+	1.801	80						
323	1.765	10						
1113	1.765	90						
229	1.758	100						
2112	1.699	80						
325	1.693	80						
2210	1.674	80						

Source: Katz *et al.* (1957).
Notes: Identification: lead hydroxide carbonate hydrate
Unit cell composition: $3PbCO_3 \cdot 2Pb(OH)_2 \cdot H_2O$
Crystalline structure: hexagonal

Table A.15.   X-Ray Powder File,
Lead Magnesium Carbonate, $PbMg(CO_3)_2$

			X-Ray Powder Data					
hkl	d (Å)	I	hkl	d (Å)	I	hkl	d (Å)	I
003	5.52	50						
101	4.13	75						
012	3.79	75						
104	2.97	100						
006	2.76	25						
015	2.62	50						
110	2.46	50						
113	2.25	50						
211	2.11	50						

Source: Lippmann (1966).
Notes: Identification: lead magnesium carbonate
Unit cell composition: $PbMg(CO_3)_2$
Crystalline structure: hexagonal (rhombohedron)

Table A.16.   X-Ray Powder File,
Lead Metaborate Hydrate, $Pb(BO_2)_2 \cdot H_2O$

			X-Ray Powder Data					
hkl	d (Å)	I	hkl	d (Å)	I	hkl	d (Å)	I
	4.43	50		1.850	50			
	4.20	50		1.694	50			
	3.59	90		1.649	10			
	3.26	100		1.609	20			
	2.70	20		1.581	40			
	2.62	100		1.528	30			
	2.48	30		1.511	10			
	2.22	50		1.483	20			
	2.11	40						
	2.09	40						
	2.04	40						
	1.879	30						

Source: Polytechnical Institute of Brooklyn (1965).
Notes: Identification: lead metaborate hydrate
Unit cell composition: $Pb(BO_2)_2 \cdot H_2O$
Crystalline structure:

Table A.17. X-Ray Powder File, Lead Oxide,
PbO II, Yellow

hkl	d (Å)	I	hkl	d (Å)	I	hkl	d (Å)	I
001	5.893	6	220	1.797	14	204	1.297	2
111	3.067	100	113	1.724	15	313	1.289	3
002	2.946	31	311	1.640	13	024	1.252	2
200	2.744	28	203	1.596	<1	402	1.244	2
201	2.493	<1	222	1.534	9	133	1.203	4
020	2.377	20	213	1.514	2	040	1.188	3
112	2.278	<1	131	1.474	11	420	1.188	3
211	2.203	<1	004	1.474	11	331	1.174	4
202	2.008	12	321	1.408	<1	224	1.139	2
003	1.963	2	400	1.372	1	115	1.120	2
212	1.850	14	114	1.363	<1	042	1.102	4
022	1.850	14	223	1.325	1	422	1.102	4
240	1.091	2						

Source: Swanson *et al*. (1951).
Notes: Identification: lead (II) oxide
Unit cell composition: PbO (yellow)
Crystalline structure: orthorhombic

Table A.18. X-Ray Powder File, Lead Oxide,
Alpha-PbO, Red, Orthorhombic

hkl	d (Å)	I	hkl	d (Å)	I	hkl	d (Å)	I
				X-Ray Powder Data				
001	4.9893	3	311	1.6710	16	041	1.3487	<1
111	3.1037	100	131	1.6699	16	331	1.2773	6
200	2.8042	16	003	1.6631	<1	223	1.2740	<1
020	2.8018	16	222	1.5518	11	420	1.2539	3
002	2.4946	11	113	1.5336	11	240	1.2532	3
201	2.4446	<1	312	1.4454	<1	004	1.2473	1
021	2.4430	<1	132	1.4447	<1	402	1.2223	2
112	2.1113	1	203	1.4304	<1	042	1.2215	2
220	1.9820	16	023	1.4301	<1	421	1.2160	<1
202	1.8639	13	400	1.4021	2	241	1.2155	<1
022	1.8632	13	040	1.4009	2	313	1.2131	5
221	1.8420	<1	401	1.3498	<1	133	1.2127	5
114	1.1898	<1						
332	1.1677	<1						
204	1.1397	2						
024	1.1395	2						
422	1.1203	4						
242	1.1198	4						

Source: Boher (1984); Boher *et al.* (1984).
Notes: Identification: lead oxide
Unit cell composition: alpha-PbO
Crystalline structure: orthorhombic

Table A.19. X-Ray Powder File, Lead Oxide,
PbO, Monoclinic

			X-Ray Powder Data					
hkl	d (Å)	I	hkl	d (Å)	I	hkl	d (Å)	I
110	7.87	40	131	3.144	100	331	2.496	20b
200	5.74	20	320	3.127	100	232	2.408	20
210	5.07	60	122	3.044	20	241	2.331	20
211	4.21	20	122	3.021	20	123	2.276	20
121	4.15	20	400	2.872	20	430	2.248	40
002	3.86	20	222	2.774	40	412	2.235	20
300	3.82	20	302	2.745	80	242	2.074	20
012	3.63	40	040	2.708	100	133	2.059	20
221	3.53	20	302	2.699	60b	323	1.982	40
130	3.45	20	140	2.636	20	342	1.928	80b
311	3.297	20	411	2.598	20	104	1.911	80b
031	3.268	20	420	2.539	20	621	1.765	20
532	1.742	20	470	1.363	20			
532	1.725	20						
352	1.702	20						
261	1.682	20						
324	1.654	60						
622	1.646	60						
360	1.633	80b						
162	1.623	60						
452	1.584	20						
262	1.572	60						
144	1.563	60						
821	1.372	20						

Source: Bosquet *et al.* (1962).
Notes: Identification: lead oxide
Unit cell composition: $PbO_{1.570}$
Crystalline structure: monoclinic

Table A.20. X-Ray Powder File,
Lead Oxide, PbO, Tetragonal

| | | | | X-Ray Powder Data | | | | |
hkl	d (Å)	I	hkl	d (Å)	I	hkl	d (Å)	I
111	3.15	100	024	1.224	12			
002	2.745	50	402	1.214	12			
200	2.707	50	420	1.211	12			
202	1.928	45	224	1.115	6			
220	1.915	45	422	1.108	6			
113	1.651	45	115	1.055	7			
311	1.635	45	333	1.048	7			
222	1.572	12	511	1.043	7			
004	1.372	5						
400	1.354	5						
133	1.250	15						
331	1.243	15						

Source: Sorrell (1973).
Notes: Identification: lead oxide
Unit cell composition: $PbO_{1.55}$
Crystalline structure: tetragonal

Table A.21. X-Ray Powder File,
Lead Oxide, PbO, Cubic

| | | | | X-Ray Powder Data | | | | |
hkl	d (Å)	I	hkl	d (Å)	I	hkl	d (Å)	I
111	3.16	100						
200	2.734	80						
222	1.934	65						
311	1.649	70						
220	1.579	11						
400	1.367	8						
331	1.255	14						
420	1.223	17						
422	1.116	8						
511	1.052	11						

Source: Sorrell (1973).
Notes: Identification: lead oxide
Unit cell composition: $PbO_{1.44}$
Crystalline structure: cubic

Table A.22.   X-Ray Powder File, Lead Oxide,
PbO, Hexagonal

				X-Ray Powder Data				
hkl	d (Å)	I	hkl	d (Å)	I	hkl	d (Å)	I
100	9.1	10	600	1.520	10			
101	6.76	10						
002	5.09	10						
300	3.04	100						
113	2.861	80						
004	2.564	30						
221	2.540	10						
410	1.990	10						
411	1.950	50						
501	1.790	50						
006	1.705	30						
324	1.620	30						

Source: Sorrell (1973).
Notes: Identification: lead oxide
Unit cell composition: $PbO_{1.37}$
Crystalline structure: hexagonal

Table A.23.   X-Ray Powder File, Lead Oxide, PbO_2, Black

				X-Ray Powder Data				
hkl	d (Å)	I	hkl	d (Å)	I	hkl	d (Å)	I
110	3.50	100	321	1.274	20	511	0.934	8
101	2.79	95	400	1.240	8	422	0.927	10
200	2.47	40	222	1.218	12	521	0.8870	12
111	2.44	6	330	1.169	8	440	0.8870	6
210	2.22	6	312	1.151	14	323	0.8720	10
211	1.855	80	411	1.133	14	530	0.8510	6
220	1.752	18	420	1.109	10	512	0.8420	8
002	1.692	14	103	1.102	10	114	0.8230	20
310	1.568	20	213	1.005	12	601	0.8020	8
112	1.524	25	510	0.972	6		0.7930	12
301	1.486	20	332	0.961	14		0.7840	10
202	1.398	14	431	0.951	16	442	0.7780	12

Source: White (1970).
Notes: Identification: lead oxide
Unit cell composition: PbO_2
Crystalline structure: tetragonal

Table A.24.   X-Ray Powder File, Lead Oxide,
Beta, PbO_2, White, Tetragonal

hkl	d (Å)	I	hkl	d (Å)	I	hkl	d (Å)	I
			X-Ray Powder Data					
110	3.51	100	222	1.218	9	[323]	0.873	9
011	2.80	95	330	1.168	4	[004]	0.8453	14
020	2.48	30	132	1.150	12	[413]	0.8230	18
121	1.855	70	141	1.133	14			
220	1.755	16	240	1.110	12			
002	1.690	7	013	1.100	12			
130	1.570	16	123	1.005	14			
112	1.525	20	150	0.972	4			
031	1.490	18	332	0.961	5			
022	1.398	10	051	0.952	12			
231	1.275	14	[422]	0.928	16			
040	1.240	5	251	0.889	9			

Source: Harada *et al.* (1981).
Notes: Identification: lead oxide
Unit cell composition: Beta-PbO_2
Crystalline structure: tetragonal

Table A.25.   X-Ray Powder File, Lead Oxide,
Pb_2O_3, Monoclinic

hkl	d (Å)	I	hkl	d (Å)	I	hkl	d (Å)	I
			X-Ray Powder Data					
001	6.953	3	402	1.9363	3	422	1.5948	3
011	4.373	2	202	1.8819	5	222	1.5642	6
002	3.4753	8	412	1.8308	19	410	1.5426	7
200	3.2075	50	404	1.7951	9			
212	3.0263	100	212	1.7849	12			
012	2.9568	92	004	1.7378	6			
020	2.8124	38	224	1.6903	15			
210	2.7868	15	232	1.6620	17			
203	2.7750	2	014	1.6603	6			
022	2.1862	4	032	1.6502	13			
204	2.1152	17	400	1.6039	3			
220	2.1147	33	215	1.6009	2			

Source: Hill (1983).
Notes: Identification: lead oxide
Unit cell composition: Pb_2O_3
Crystalline structure: monoclinic

Table A.26.   X-Ray Powder File, Lead Oxide,
Pb_3O_4, Red, Tetragonal

| | | | X-Ray Powder Data | | | | | |
hkl	d (Å)	I	hkl	d (Å)	I	hkl	d (Å)	I
110	6.23	12	330	2.076	2	521	1.5876	12
201	3.66	4	411	2.032	12	114	1.5876	12
211	3.38	100	420	1.970	12	440	1.5580	3
002	3.28	8	213	1.903	20	512	1.5292	8
220	3.113	20	421	1.887	<1	413	1.5292	8
112	2.903	50	402	1.829	20	530	1.5116	2
310	2.787	45	332	1.755	30	531	1.4744	<1
202	2.632	30	510	1.729	2	600	1.4687	4
320	2.444	2	431	1.7025	2	224	1.4521	2
321	2.289	4	422	1.6897	2	314	1.4144	14
222	2.260	8	004	1.6417	8	611	1.4144	14
400	2.205	2	323	1.6302	<1	620	1.3944	<1
532	1.3728	2						
541	1.3471	4						
523	1.3109	4						
622	1.2830	4						
424	1.2614	4						
710	1.2461	4						
215	1.2461	4						

Source: *NBS Circular* (1958).
Notes: Identification: lead (II, III) oxide
Unit cell composition: Pb_3O_4
Crystalline structure: tetragonal

Table A.27.   X-Ray Powder File, Lead Oxide,
$Pb_{12}O_{19}$, Brown Black, Monoclinic

hkl	d (Å)	I	hkl	d (Å)	I	hkl	d (Å)	I
220	3.27	100	226	1.650	30	463	1.244	6
023	3.13	90	260	1.638	45	029	1.244	6
203	2.750	40	063	1.635	50	446	1.226	2
040	2.711	50	423	1.635	50	283	1.216	4
203	2.698	40	226	1.627	25	283	1.212	2
015	2.249	4	440	1.575	30	603	1.208	4
034	2.249	4	046	1.566	25	209	1.208	4
400	1.932	40	406	1.374	4	643	1.119	2
243	1.932	40	080	1.356	4	480	1.110	4
006	1.917	30	406	1.348	2	086	1.107	4
243	1.912	40	620	1.254	8	249	1.103	2
423	1.658	35	463	1.254	8	626	1.059	2
	1.049	2?						
2$\overline{100}$	1.044	4						
0$\overline{103}$	1.044	4						
$\overline{626}$	1.038	<2						
429	1.038	<2						
	0.965	2						

Source: White *et al.* (1964).
Notes: Identification: lead oxide
Unit cell composition: $Pb_{12}O_{19}$
Crystalline structure: monoclinic

Table A.28. X-Ray Powder File, Lead Oxide Carbonate, $PbCo_3 \cdot PbO$, Pb_2CO_4

				X-Ray Powder Data				
hkl	d (Å)	I	hkl	d (Å)	I	hkl	d (Å)	I
	6.48	20		2.498	6b		1.841	4b
	4.14	30		2.471	4b		1.822	10
	4.02	55		2.328	20		1.811	10b
	3.45	2		2.288	30		1.801	4b
	3.39	16		2.253	35		1.750	16
	3.22	100		2.183	35		1.731	25
	3.18	95		2.160	16		1.720	16
	2.935	30		2.063	16		1.694	16
	2.859	60		2.013	25		1.667	8
	2.737	4		1.933	16		1.654	16
	2.652	20		1.910	35		1.632	8
	2.568	40		1.887	8b		1.619	6
	1.608	6		1.348	6		1.200	10
	1.597	30		1.332	6		1.196	8
	1.580	16		1.316	6		1.172	8
	1.558	4		1.308	4		1.157	4
	1.536	4		1.295	6		1.147	4
	1.499	6		1.285	6		1.141	6
	1.486	10		1.274	20		1.116	8
	1.467	10		1.261	6		1.101	6
	1.441	4		1.253	8		1.093	4
	1.423	6		1.248	8		1.079	2
	1.382	6		1.240	10		1.070	4
	1.370	4		1.225	4		1.062	8
	1.050	6						
	1.037	4						
	1.024	4						
	1.015	4						
	0.988	4						
	0.977	6						
	0.967	4						

Source: Grisafe *et al.* (1964).
Notes: Identification: lead oxide carbonate
Unit cell composition: Pb_2CO_4
Crystalline structure: orthorhombic

Table A.29. X-Ray Powder File, Lead Oxide Carbonate, $Pb_7C_4O_{15}$, $4PbCO_3 \cdot 3PbO$

| | | | | X-Ray Powder Data | | | | |
hkl	d (Å)	I	hkl	d (Å)	I	hkl	d (Å)	I
	7.12	8		3.35	4		2.614	6
	7.03	20		3.29	50		2.599	10
	6.99	20		3.25	20		2.584	16
	6.19	8		3.24	20		2.553	8
	5.72	10		3.22	35		2.503	4
	5.17	8		2.994	100		2.493	4
	4.71	10		2.962	16		2.466	2
	4.62	4		2.862	50		2.447	4
	4.42	55		2.810	35		2.361	6
	3.82	6		2.670	10		2.338	4
	3.64	16		2.664	16		2.312	4
	3.54	16		2.647	8		2.267	20
	2.201	4		1.958	10		1.724	6
	2.194	4		1.952	10		1.722	10
	2.185	8		1.949	16		1.718	10
	2.154	20		1.919	2		1.695	2
	2.148	20		1.906	8b		1.686	4
	2.140	16		1.871	8		1.676	10
	2.075	20		1.858	2		1.662	16
	2.070	20		1.852	4		1.658	16
	2.055	4		1.839	4b		1.652	16
	2.024	6		1.769	25		1.640	4
	2.018	6		1.753	20		1.625	6
	1.993	4		1.732	6		1.611	6
	1.605	4		1.396	4		1.260	4
	1.602	6		1.393	4b		1.250	4
	1.548	6		1.384	6		1.248	4
	1.561	4b		1.344	2		1.233	6
	1.543	2b		1.323	4		1.181	2
	1.499	10		1.322+	2b		1.157	4
	1.495	20		1.318	2		1.156	4
	1.464	2		1.315	4		1.146	10
	1.459	4b		1.309	4		1.125	2
	1.450	4		1.308	4		1.114	2
	1.437	4b		1.304	4		1.107	4
	1.430	6		1.301	4		1.103	4

Source: Grisafe *et al.* (1964).
Notes: Identification: lead oxide carbonate
Unit cell composition: $Pb_7C_4O_{15}$
Crystalline structure:

Table A.30. X-Ray Powder File, Lead Oxide Carbonate, $Pb_7C_4O_{15}$, $4PbCO_3 \cdot 3PbO$

hkl	d (A)	I	hkl	d (A)	I	hkl	d (A)	I
				X-Ray Powder Data				
	7.12	8		2.810	35		2.148	20
	7.03	20		2.670	10		2.140	16
	6.99	20		2.664	16		2.075	20
	6.19	8		2.647	8		2.070	20
	5.72	10		2.614	6		2.055	4
	5.17	8		2.599	10		2.024	6
	4.71	10		2.584	16		2.018	6
	4.62	4		2.553	8		1.993	4
	4.42	55		2.503	4		1.958	10
	3.82	6		2.493	4		1.952	10
	3.64	16		2.466	2		1.949	16
	3.54	16		2.447	4		1.919	2
	3.35	4		2.361	6		1.906	8
	3.29	50		2.338	4		1.871	8
	3.25	20		2.312	4		1.858	2
	3.24	20		2.267	20		1.852	4
	3.22	35		2.201	4		1.839	4
	2.994	100		2.194	4		1.769	25
	2.962	16		2.185	8		1.753	20
	2.862	50		2.154	20		1.732	6
	1.724	6		1.459	4		1.181	2
	1.722	10		1.450	4		1.157	4
	1.718	10		1.437	4		1.156	4
	1.695	2		1.430	6		1.146	10
	1.686	4		1.396	4		1.125	2
	1.676	10		1.393	4		1.114	2
	1.662	16		1.384	6		1.107	4
	1.658	16		1.344	2		1.103	4
	1.652	16		1.323	4		1.083	8
	1.640	4		1.322	2		1.074	2
	1.625	6		1.318	2		1.056	4
	1.611	6		1.315	4		0.999	4
	1.605	4		1.309	4		0.997	4
	1.602	6		1.308	4		0.979	2
	1.548	6		1.304	4		0.944	2
	1.561	4		1.301	4			
	1.543	2		1.260	4			
	1.499	10		1.250	4			
	1.495	20		1.248	4			
	1.464	2		1.233	6			

Source: Grisafe *et al.* (1964).
Notes: Identification: lead oxide carbonate
Unit cell composition: $Pb_7C_4O_{15}$
Crystalline structure:

Table A.31.   X-Ray Powder File, Lead Oxide
Carbonate Hydroxide, $Pb_{10}(CO_3)_6(OH)_6O$,
Plumbonacrite, $6PbCO_3 \cdot 3PB(OH)_2 \cdot PbO$

| | | | | X-Ray Powder Data | | | | |
hkl	d (Å)	I	hkl	d (Å)	I	hkl	d (Å)	I
100	7.70	5	213	2.797	20	403	1.914	5
102	6.63	10	300	2.619	100	227	1.914	5
110	4.55	20	303	2.506	10	404	1.877	5
111	4.47	20	304	2.420	10	30$\overline{10}$	1.808	10
112	4.26	80	216	2.420	10	41$\overline{2}$	1.699	50
113	3.98	30	221	2.261	10	413	1.680	10
202	3.75	10	222	2.235	40	415	1.622	10
114	3.66	20	223	2.188	30	417	1.543	10
115	3.36	70	224	2.133	20	40$\overline{10}$	1.543	10
116	3.07	5	225	2.066	20	$\overline{330}$	1.512	20
211	2.953	40	226	1.994	5	329	1.512	20
212	2.891	20	402	1.942	5	334	1.469	10

Source: Olby (1966).
Notes: Identification: lead oxide carbonate hydroxide
Unit cell composition: $Pb_{10}(CO_3)_6(OH)_6O$
Crystalline structure: hexagonal

Table A.32.   X-Ray Powder File, Lead Oxide Sulfate
Pb_2OSO_4, $PbSO_4·PbO$

				X-Ray Powder Data				
hkl	d (Å)	I	hkl	d (Å)	I	hkl	d (Å)	I
001	6.366	5	312	2.867	20	313	2.1731	5
200	6.193	10	020	2.849	100	022	2.1228	5
201	5.917	20	220	2.588	10	600	2.0637	5
110	5.175	20	221	2.566	10	422	2.0517	10
111	4.431	80	511	2.4766	10	603	1.9717	50
201	3.702	30	202	2.4313	10	511	1.9102	10
111	3.699	10	401	2.4012	40	222	1.8493	10
311	3.516	20	203	2.3468	30	113	1.8482	10
310	3.342	70	601	2.2766	20	712	1.8450	10
002	3.183	5	510	2.2711	20	421	1.8360	20
112	2.963	40	221	2.2575	5	131	1.8339	20
402	2.958	20	602	2.2339	5	223	1.8114	10
404	1.7631	4	531	1.5628	4			
622	1.7580	6						
204	1.7351	4						
330	1.7254	12						
801	1.6822	4						
620	1.6712	5						
132	1.6659	13						
514	1.6513	5						
332	1.6482	2						
623	1.6213	5						
114	1.6060	4						
512	1.5742	5						

Source: Hill (1983).
Notes: Identification: lead oxide sulfate
Unit cell composition: Pb_2OSO_4
Crystalline structure: monoclinic

Table A.33. X-Ray Powder File,
Lead Oxide Sulfate Pb$_2$(SO$_4$)O

				X-Ray Powder Data				
hkl	d (Å)	I	hkl	d (Å)	I	hkl	d (Å)	I
001	6.362	25	112	2.9614	80	512	2.3428	8
200	6.193	25	402	2.9499	30	601	2.2735	5
201	5.906	16	312	2.8620	25	510	2.2696	8
110	5.173	4	020	2.8481	45	403	2.2696	8
111	4.428	25	311	2.6029	2	221	2.2579	25
201	3.7029	20	021	2.5992	2	602	2.2294	7
111	3.7029	20	220	2.5876	4	421	2.1906	3
202	3.5110	15	221	2.5641	9	313	2.1694	2
311	3.5110	15	511	2.4729	18	022	2.1211	5
310	3.3426	100	202	2.4308	20	003	2.1211	5
002	3.1811	11	401	2.4012	8	113	2.1006	1
400	3.0917	1	203	2.3428	8	600	2.0624	9
422	2.0492	25	223	1.8091	3	801	1.6802	5
513	1.9970	1	621	1.7768	2	620	1.6706	7
312	1.9798	<1	423	1.7768	2	132	1.6653	18
603	1.9675	4	131	1.7689	1	514	1.6476	7
511	1.9099	10	404	1.7599	4	332	1.6576	7
130	1.8760	1	622	1.7554	5	623	1.6185	4
222	1.8482	25	331	1.7468	3	114	1.6043	5
113	1.8482	25	204	1.7328	5	331	1.5929	1
712	1.8421	6	330	1.7247	18	512	1.5741	9
421	1.8358	9	802	1.7151	1	531	1.5618	5
131	1.8332	5	713	1.7117	1	313	1.5587	3
711	1.8332	5	710	1.6884	2	422	1.5530	1
800	1.5463	1	913	1.4381	4			
714	1.5136	6	911	1.4362	2			
530	1.5065	2	040	1.4242	9			
424	1.4971	6	405	1.4128	1			
711	1.4860	7	332	1.4128	1			
224	1.4804	4						
804	1.4754	8						
912	1.4754	8						
822	1.4692	1						
602	1.4652	<1						
114	1.4477	10						
821	1.4477	10						

Source: Mentzen *et al.* (1983).
Notes: Identification: lead oxide sulfate
Unit cell composition: Pb$_2$(SO$_4$)O
Crystalline structure: monoclinic

Table A.34. X-Ray Powder File,
Lead Oxide Sulfate, Alpha-$Pb_3O_2SO_4$

			X-Ray Powder Data					
hkl	d (Å)	I	hkl	d (Å)	I	hkl	d (Å)	I
001	7.86	80	201	2.99	100		2.136	20
101	5.89	50	202	2.95	80		2.121	20
101	4.76	200	020	2.89	80		2.075	50
110	4.46	20	021	2.716	50		2.066	50
111	4.14	20	202	2.394	50		1.968	80
102	3.79	50	301	2.394	50		1.897	50
111	3.67	50	013	2.379	50		1.836	50
200	3.50	50	203	2.357	50		1.828	80
201	3.50	50	300	2.357	50		1.796	50
012	3.26	100	122	2.299	50		1.732	80
112	3.19	20		2.232	20		1.689	80
210	2.99	100		2.206	20		1.628	20
	1.577	50						
	1.567	50						
	1.526	50						
	1.512	50						

Source: Bowin *et al.* (1968).
Notes: Identification: lead oxide sulfate
Unit cell composition: Alpha-Pb_3SO_6
Crystalline structure: monoclinic

Table A.35. X-Ray Powder File,
Lead Oxide Sulfate, Alpha-Pb_3SO_6

			X-Ray Powder Data					
hkl	d (Å)	I	hkl	d (Å)	I	hkl	d (Å)	I
001	7.844	31	020	2.8850	38	004	1.9620	8
101	5.889	9	112	2.7625	5	114	1.8934	6
101	4.744	6	021	2.7080	4	313	1.8588	3
110	4.453	6	121	2.4648	3	222	1.8318	14
111	4.1219	4	013	2.3828	10	223	1.8229	18
102	3.7871	6	202	2.3713	15	104	1.7918	11
111	3.6641	8	203	2.3516	15	401	1.7918	11
012	3.2452	100	122	2.2950	9	402	1.7437	2
210	2.9926	41	103	2.2945	4	032	1.7272	13
211	2.9848	67	311	2.2009	3	321	1.7063	3
102	2.9683	13	301	2.1162	4	230	1.6857	6
202	2.9453	13	221	2.0691	17	231	1.6842	11
024	1.6223	6						
411	1.5732	9						
413	1.5640	10						
033	1.5496	2						

Source: Hill (1983).
Notes: Identification: lead oxide sulfate
Unit cell composition: Alpha-$Pb_3O_2SO_4$
Crystalline structure: monoclinic

Table A.36.   X-Ray Powder File, Lead Sulfate, $PbSO_4$

				X-Ray Powder Data				
hkl	d (Å)	I	hkl	d (Å)	I	hkl	d (Å)	I
110	5.381	2	112	2.4131	5	222	1.9051	5
101	4.267	50	221	2.4079	20	132	1.8793	8
020	4.238	51	131	2.3557	2	330	1.7928	19
111	3.811	40	022	2.2768	20	103	1.7421	8
120	3.620	15	310	2.2374	5	312	1.7226	2
200	3.480	22	230	2.1937	7	241	1.7161	4
021	3.334	67	122	2.1637	30	410	1.7041	19
210	3.219	53	202	2.1328	7	331	1.7013	16
121	3.008	88	212	2.0675	100	023	1.6562	9
211	2.765	40	231	2.0323	58	150	1.6472	4
002	2.700	43	140	2.0272	54	411	1.6251	7
130	2.619	10	041	1.9727	24	142	1.6210	22
123	1.6113	12	060	1.4123	<1	511	1.3306	2
151	1.5753	2	152	1.4060	3	351	1.3267	3
213	1.5703	7	313	1.4020	15	520	1.3220	3
421	1.5420	1	233	1.3911	8	441	1.3044	3
250	1.5240	1	160	1.3845	4	024	1.2859	6
332	1.4932	23	422	1.3818	2	521	1.2840	3
133	1.4826	1	043	1.3717	8	243	1.2757	1
251	1.4667	8	350	1.3687	14			
402	1.4620	7	061	1.3666	<1			
412	1.4405	12	004	1.3496	8			
431	1.4286	8	501	1.3472	7			
303	1.4219	<1	161	1.3410	8			

Source: McMurdie *et al.* (1986).
Notes: Identification: lead sulfate
Unit cell composition: $PbSO_4$
Crystalline structure: orthorhombic

Table A.37. X-Ray Powder File, Lead Silicate, Pb_3SiO_5

			X-Ray Powder Data					
hkl	d (Å)	I	hkl	d (Å)	I	hkl	d (Å)	I
011	7.01	6	031	3.245	18	411	2.431	1
111	6.73	6	113	3.202	6	304	2.406	2
111	5.41	6	202	3.175	8	233	2.373	6
020	5.17	6	122	3.144	100	314	2.344	4
211	5.04	1	213	3.119	9	431	2.260	9
002	4.78	5	322	2.946	90	104	2.196	1
202	4.44	2	400	2.832	30	324	2.181	2
211	3.965	7	113	2.733	12	402	2.146	1
311	3.716	10	231	2.688	4	342	2.098	1
302	3.585	5	040	2.586	20	520	2.074	1
022	3.510	4	204	2.516	35	332	2.065	2
222	3.367	4	420	2.483	5	431	2.024	10
504	2.003	3	525	1.6787				
204	1.9813	9	144	1.6742				
034	1.9643	3						
143	1.9098	12						
613	1.8902	1						
351	1.8357	4						
244	1.8030	11						
604	1.7938	9						
243	1.7800	2						
235	1.7449	5						
632	1.7242	7						
624	1.6945	5						

Source: Breuer *et al*. (1980).
Notes: Identification: lead silicate
Unit cell composition: Pb_3SiO_5
Crystalline structure: monoclinic

Table A.38.   X-Ray Powder File, Lead Silicate, $PbSiO_3$

| | | | | X-Ray Powder Data | | | | |
hkl	d (Å)	I	hkl	d (Å)	I	hkl	d (Å)	I
101	6.45	25	113	3.30	45	204	2.800	55
011	5.86	55	303	3.25	60	113	2.747	30
111	5.72	55	202	3.23	70	221	2.735	45
002	5.18	20	301	3.14	10	321	2.670	35
210	4.41	5	121	3.10	20	322	2.595	20
012	4.18	15	221	3.05	50	223	2.532	30
301	4.08	10	402	3.02	45	023	2.464	10
103	3.73	20	122	2.987	70	404	2.441	5
112	3.56	95	022	2.911	40	501	2.416	5
020	3.53	75	311	2.883	30	213	2.380	5
312	3.37	45	222	2.860	10	321	2.347	35
021	3.34	100	400	2.815	60	421	2.300	75
114	2.206	20						
215	2.140	45						
204	2.060	20						
602	2.040	25						
421	2.028	45						

Source: Medenback *et al.* (1975).
Notes: Identification: lead silicate
Unit cell composition: $PbSiO_3$
Crystalline structure: monoclinic

Appendix B

Infrared Spectra of Resins and Binders

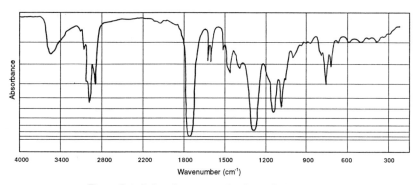

Figure B.1. Infrared spectrum of a short oil alkyd resin.

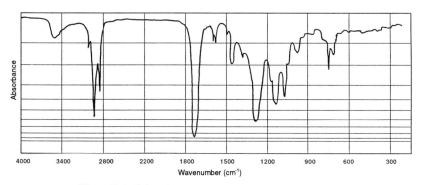

Figure B.2. Infrared spectrum of a medium oil alkyd resin.

255

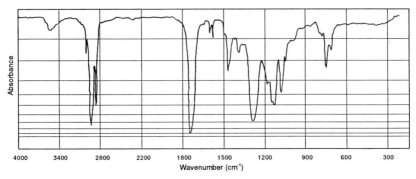

Figure B.3. Infrared spectrum on a long oil alkyd.

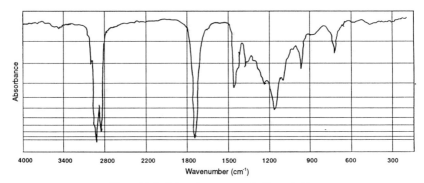

Figure B.4. Infrared spectrum of a linseed oil.

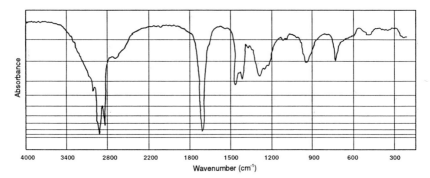

Figure B.5. Infrared spectrum of a soya oil.

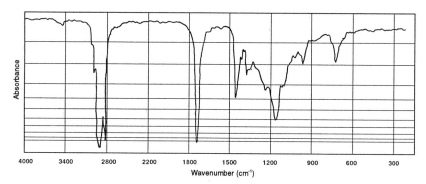

Figure B.6. Infrared spectrum of a fish oil.

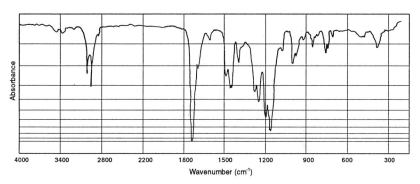

Figure B.7. Infrared spectrum of an acrylic resin.

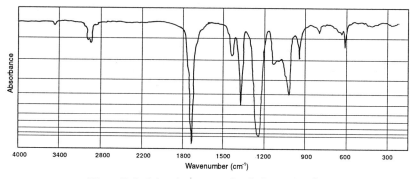

Figure B.8. Infrared spectrum of a vinyl acetate resin.

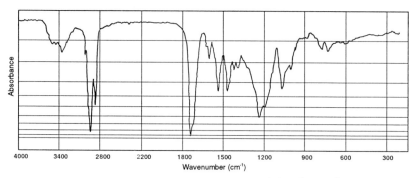

Figure B.9. Infrared spectrum of an oil-modified urethane resin.

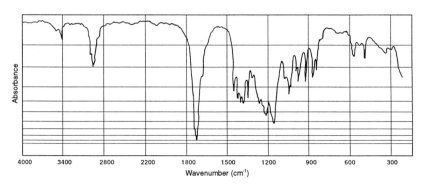

Figure B.10. Infrared spectrum of a saturated polyester resin.

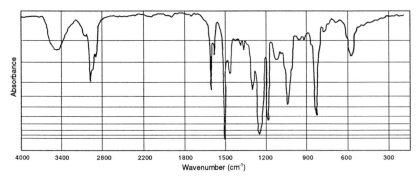

Figure B.11. Infrared spectrum of an epoxy resin.

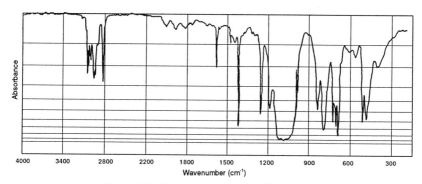

Figure B.12. Infrared spectrum of a silicone resin.

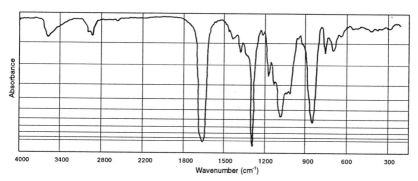

Figure B.13. Infrared spectrum of a nitrocellulose resin.

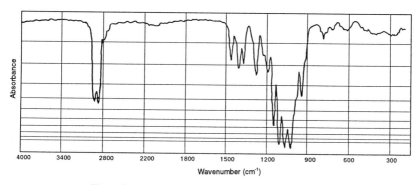

Figure B.14. Infrared spectrum of a polysulfide resin.

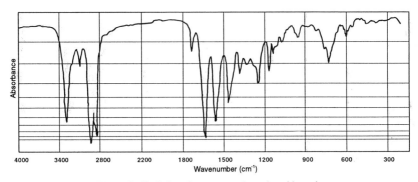

Figure B.15. Infrared spectrum of a polyamide resin.

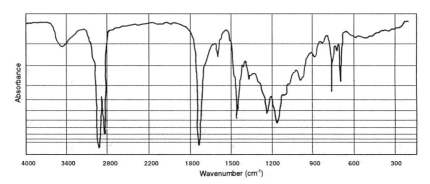

Figure B.16. Infrared spectrum of a phenolic oil resin.

Appendix C

Federal and State Hazardous Waste Contacts

State Hazardous Waste Contacts

Alabama

Land Division
Alabama Department of Environmental Management
1751 Federal Drive
Montgomery, Alabama 36130
(205) 271-7730

Alaska

U.S. Environmental Protection Agency—Region X
Waste Management Branch
MS 11W-112
1200 Sixth Avenue
Seattle, Washington 98101
(206) 442-0151

American Samoa

Environmental Quality Commission
Government of American Samoa
Pago Pago, American Samoa 96799
Overseas Operator Commercial call: Country code (684) 663-2304

Arizona

Office of Waste and Water Quality Management
Arizona Department of Environmental Quality
2005 North Central Avenue, Room 304
Phoenix, Arizona 85004
(602) 257-2305

Arkansas

Arkansas Department of Pollution Control and Ecology
P.O. Box 9583
Little Rock, Arkansas 72219
(501) 562-7444

California

California Department of Health Services
Toxic Substance Control Division
Department of Health Services
P.O. Box 942732, 400 P. Street
Sacramento, California 95814
(916) 323-2913

Colorado

Hazardous Materials and Waste Management Division
Colorado Department of Health
4210 East 11th Avenue
Denver, Colorado 80220
(303) 331-4830

Connecticut

Waste Management Bureau
Department of Environmental Protection
State Office Building
Hartford, Connecticut 06106
(203) 566-8844

Delaware

Delaware Department of Natural Resources and Environmental Control
Division of Air and Waste Management
Hazardous Waste Management Branch
P.O. Box 1401, 89 Kings Highway
Dover, Delaware 19903
(302) 736-3689

District of Columbia

Department of Consumer and Regulatory Affairs
Environmental Control Division
Pesticides and Hazardous Waste Branch
2100 Martin Luther King, Jr., Avenue, South East
Room 204
Washington, DC 20020

Florida

Hazardous Waste Section
Department of Environmental Regulations
Twin Towers Office Building
2600 Stone Road
Tallahassee, Florida 32399-2400
(904) 488-0300

Georgia

Land Protection Branch
Industrial and Hazardous Waste Management Program
Floyd Towers East, 205 Butler Street, South East
Atlanta, Georgia 30334
(404) 656-2833

Guam

Guam Environmental Protection Agency
IT&E
Harmon Plaza Complex, Unit D-107
130 Rojas Street
Harmon, Guam 96911
Overseas Operator Commercial call: Country code (671) 646-7579

Hawaii

For information or forms, contact:
Department of Health
Hazardous Waste Program
P.O. Box 3378
Honolulu, Hawaii 96801
(808) 548-2270

Send completed forms to:
U.S. Environmental Protection Agency—Region IX
Resource Conservation and Recovery Act Programs Section
Hazardous Waste Management Division
1235 Mission Street
San Francisco, California 94103

Idaho

Idaho Department of Health and Welfare
Tower Building, Third Floor
450 West State Street
Boise, Idaho 83720
(208) 334-5879

Illinois

For information or forms, contact:
U.S. Environmental Protection Agency—Region V
Resource Conservation and Recovery Act Activities
Waste Management Division
P.O. Box A3587
Chicago, Illinois 60690
(312) 886-4001

Send completed forms to:
Illinois Environmental Protection Agency
Division of Land Pollution Control
2200 Churchill Road
Springfield, Illinois 62706
(217) 782-6760

Indiana

For information or forms, contact:
Indiana Department of Environmental Management
105 South Meridian Street
P.O. Box 6015
Indianapolis, Indiana 46225
(317) 232-3210

Send completed forms to:
U.S. Environmental Protection Agency—Region V
Resource Conservation and Recovery Act Activities
Waste Management Division
P.O. Box A3587
Chicago, Illinois 60690

Iowa

U.S. Environmental Protection Agency—Region VII
Resource Conservation and Recovery Act Branch
726 Minnesota Avenue
Kansas City, Kansas 66101
(913) 236-2852 or 1 (800) 223-0425

Kansas

Bureau of Air and Waste Management
Department of Health and Environment
Forbes Field, Building 740
Topeka, Kansas 66620
(913) 296-1600

Kentucky

Division of Waste Management
Department of Environmental Protection
Cabinet for Natural Resources and Environmental Protection
Fort Boone Plaza, Building 1
Frankfort, Kentucky 40601
(502) 564-6716

Louisiana

Louisiana Department of Environmental Quality
Department of Solid and Hazardous Waste
P.O. Box 44307
Baton Rouge, Louisiana 70804
(504) 342-1345

Maine

Bureau of Oil and Hazardous Materials Control
Department of Environmental Protection
Ray Building, Station 17
Augusta, Maine 04333
(207) 289-2651

Maryland

Maryland Department of the Environment
Waste Management Administration
2500 Broening Highway
Baltimore, Maryland 21224
(301) 631-3304

Massachusetts

Division of Hazardous Waste
Department of Environmental Protection
One Winter Street, Fifth Floor
Boston, Massachusetts 02108
(617) 292-5851

Michigan

For information or forms, contact:
Waste Management Division
Environmental Protection Bureau
Department of Natural Resources
Box 30038
Lansing, Michigan 48909
(517) 373-2730

Send completed forms to:
U.S. Environmental Protection Agency—Region V
Resource Conservation and Recovery Act Activities
Waste Management Division
P.O. Box A3587
Chicago, Illinois 60690

Minnesota

For information or forms, contact:
Solid and Hazardous Waste Division
Minnesota Pollution Control Agency
520 Lafayette Road, North
St. Paul, Minnesota 55155
(612) 296-7282

Send completed forms to:
U.S. Environmental Protection Agency—Region V
Resource Conservation and Recovery Act Activities
Waste Management Division
P.O. Box A3587
Chicago, Illinois 60690

Mississippi

Hazardous Waste Division
Bureau of Pollution Control
Department of Environmental Quality
P.O. Box 10385
Jackson, Mississippi 39289-0385
(601) 961-5062

Missouri

Waste Management Program
Department of Natural Resource
Jefferson Building
205 Jefferson Street (13/14 Floor)
P.O. Box 176
Jefferson City, Missouri 65102
(314) 751-3176

Montana

Solid and Hazardous Waste Bureau
Department of Health and Environmental Sciences
Cogswell Building
Helena, Montana 59620
(406) 444-1430

Nebraska

Hazardous Waste Management Section
Department of Environmental Control
State House Station
P.O. Box 98922
Lincoln, Nebraska 68509-8922
(402) 471-4217

Nevada

Waste Management Bureau
Division of Environmental Protection
Department of Conservation and Natural Resources
Capital Complex
123 West Nye Lane
Carson City, Nevada 89710
(702) 687-5872

New Hampshire

Department of Environmental Services
Waste Management Division
6 Hazen Drive
Concord, New Hampshire 03301
(603) 271-2900

New Jersey

For information contact:
New Jersey Department of Environmental Protection
Division of Waste Management
Bureau of Hazardous Waste Classification and Manifests
401 East State Street, CN-028
Trenton, New Jersey 08625
(609) 292-8341

Obtain forms and send completed forms to:
U.S. Environmental Protection Agency—Region II
Permits Administration Branch
26 Federal Plaza, Room 505
New York, New York 10278

New Mexico

New Mexico Health and Environmental Department
Hazardous Waste Bureau
1190 St. Francis Drive
Santa Fe, New Mexico 87503
(505) 827-2929

New York

For information contact:
New York Department of Environmental Conservation
Division of Hazardous Waste Substance Regulation
P.O. Box 12820
Albany, New York 12212
(518) 457-0530

Obtain forms and send completed forms to:
U.S. Environmental Protection Agency—Region II
Permits Administration Branch
26 Federal Plaza, Room 505
New York, New York 10278

North Carolina

Hazardous Waste Section
Division of Solid Waste Management
Department of Environment, Health, and Natural Resources
P.O. Box 27687
Raleigh, North Carolina 27611-7687
(919) 733-2178

North Dakota

Division of Waste Management
Department of Health and Consolidated Laboratories
1200 Missouri Avenue
P.O. Box 5520
Bismarck, North Dakota 58502-5520
(701) 224-2366

Northern Mariana Islands

For information or forms, contact:
Department of Public Health and Environmental Services
Division of Environmental Quality
Dr. Torres Hospital
P.O. Box 1304
Siapan, Mariana Islands 96950
Overseas Operator Commercial call: Country code (676) 234-6984

Send completed forms to:
U.S. Environmental Protection Agency—Region IX
Resource Conservation and Recovery Act Programs Section (H-2-3)
Hazardous Waste Management Division
1235 Mission Street
San Francisco, California 94103

Ohio

U.S. Environmental Protection Agency—Region V
Resource Conservation and Recovery Act Activities
Waste Management Division
P.O. Box A3587
Chicago, Illinois 60690
(312) 886-4001

Oklahoma

Oklahoma State Department of Health
Industrial Waste Division
1000 Northeast 10th Street
Oklahoma City, Oklahoma 73152
(405) 271-5338

Oregon

Oregon Department of Environmental Quality
Hazardous Waste Operations
811 Southwest 6th Avenue
Portland, Oregon 97204
(503) 229-5913

Pennsylvania

For information or forms, contact:
Pennsylvania Department of Environmental Resources
Bureau of Waste Management
P.O. Box 2063
Harrisburg, Pennsylvania 17120
(717) 787-9870

Send completed forms to:
U.S. Environmental Protection Agency—Region III
Resource Conservation and Recovery Act Programs Branch
Pennsylvania Section (3 IIW51)
841 Chestnut Building
Philadelphia, PA 19107

Puerto Rico

For information or forms, contact:
Puerto Rico Environmental Quality Board
Land Pollution Control Area
Inspection, Monitoring, and Surveillance
P.O. Box 11488
Santurce, Puerto Rico 00910-1488
(809) 722-0439

Obtain forms and send completed forms to:
U.S. Environmental Protection Agency—Region II
Permits Administration Branch
26 Federal Plaza, Room 505
New York, New York 10278

Rhode Island

Division of Air and Hazardous Materials
Department of Environmental Management
291 Promenade Street
Providence, Rhode Island 02908-5767
(401) 277-2808

South Carolina

Bureau of Solid Waste Management
Hazardous Waste Management
Department of Health and Environmental Control
2600 Bull Street
Columbia, South Carolina
(803) 734-2500

South Dakota

Office of Waste Management
Department of Water and Natural Resources
Joe Foss Building, 523 East Capital Street
Pierre, South Dakota 57501-3181
(605) 773-3153

Tennessee

Division of Solid Waste Management
Tennessee Department of Health and Environment
701 Broadway, Customs House, Fourth Floor
Nashville, Tennessee 37247-3530

Texas

Texas Water Commission
Compliance Assistance Unit
Hazardous and Solid Waste Division
P.O. Box 13087, Capital Station
Austin, Texas 78711-3087
(512) 463-8175

Utah

Bureau of Solid and Hazardous Waste Management
Department of Health
P.O. Box 16690, 288 North 1460 West
Salt Lake City, Utah 84116-0690
(801) 538-6170

Vermont

Hazardous Materials Management Division
Department of Environmental Conservation
103 South Main Street
Waterbury, Vermont 05676

Virginia

Virginia Department of Waste Management
Monroe Building, Eleventh Floor
101 North 14th Street
Richmond, Virginia 23219
(804) 225-2667

Virgin Islands

For information or forms, contact:
Virgin Islands Department of Planning and Natural Resources
Division of Environmental Protection
179 Altona and Welgunst
St. Thomas, Virgin Islands 00801
(809) 774-3320

Obtain forms and send completed forms to:
U.S. Environmental Protection Agency—Region II
Permits Administration Branch
26 Federal Plaza, Room 505
New York, New York 10278

Washington

Solid and Hazardous Waste Management Division
Department of Ecology, Mail Stop PV-11
Olympia, Washington 98504
(206) 459-6369

West Virginia

West Virginia Division of Natural Resources
Waste Management Section
1356 Hansford Street
Charleston, West Virginia 25301
(304) 348-5393

Wisconsin

For information or forms, contact:
Bureau of Solid Waste
Department of Natural Resources
P.O. Box 7921
Madison, Wisconsin 53707
(608) 266-1327

Send completed forms to:
U.S. Environmental Protection Agency—Region V
Resource Conservation and Recovery Act Activities
Waste Management Division
P.O. Box A3587
Chicago, Illinois 60690

Wyoming

U.S. Environmental Protection Agency—Region VIII
Hazardous Waste Management Division (8HWM-ON)
999 18th Street, Suite 500
Denver, Colorado 80202-2405
(303) 293-1795

U.S. ENVIRONMENTAL PROTECTION AGENCY REGIONAL OFFICES

I   Connecticut, Maine, Massachusetts, New Hampshire, Rhode Island, Vermont

U.S. Environmental Protection Agency—Region I
Waste Management Division
JFK Federal Building
Boston, Massachusetts 02203-2211

II   New Jersey, New York, Puerto Rico, Virgin Islands

U.S. Environmental Protection Agency—Region II
Permits Administration Branch
26 Federal Plaza, Room 505
New York, New York 10278

III   Delaware, District of Columbia, Maryland, Pennsylvania, Virginia, West Virginia

U.S. Environmental Protection Agency—Region III
Resource Conservation and Recovery Act Programs Branch (3 HW 53)
841 Chestnut Street
Philadelphia, Pennsylvania 19107

IV   Alabama, Florida, Georgia, Kentucky, Mississippi, North Carolina, South Carolina, Tennessee

U.S. Environmental Protection Agency—Region IV
Hazardous Waste Management Division
345 Courtland Street, North East
Atlanta, Georgia 30332

V   Illinois, Indiana, Michigan, Minnesota, Ohio, Wisconsin

U.S. Environmental Protection Agency—Region V
Resource Conservation and Recovery Act Activities
Waste Management Division
P.O. Box A3587
Chicago, Illinois 60690

VI   Arkansas, Louisiana, New Mexico, Oklahoma, Texas

U.S. Environmental Protection Agency—Region VI
Hazardous Waste Management Division
First Interstate Bank Tower
1445 Ross Avenue, Suite 1200
Dallas, Texas 75202-2733

VII   Iowa, Kansas, Nebraska, Missouri

U.S. Environmental Protection Agency—Region VII
Resource Conservation and Recovery Act Branch
726 Minnesota Avenue
Kansas City, Kansas 66620

VIII   Colorado, Montana, North Dakota, South Dakota, Utah, Wyoming

U.S. Environmental Protection Agency—Region VIII
Hazardous Waste Management Division (8HWM-ON)
999 18th Street, Suite 500
Denver, Colorado 80202-2405

IX   Arizona, California, Hawaii, Nevada, American Samoa, Guam,
Northern Mariana Islands

U.S. Environmental Protection Agency—Region IX
Hazardous Waste Management Division
1235 Mission Street
San Francisco, California 94103

X   Alaska, Idaho, Oregon, Washington

U.S. Environmental Protection Agency—Region X
Waste Management Branch (HW-112)
1200 6th Avenue
Seattle, Washington 98101

References

3M Corporation. 1990. "Safest Stripper and Varnish Remover." Technical bulletin.

AAP (American Academy of Pediatrics) Subcommittee on Accidental Poisoning. 1969. "Prevention, diagnosis, and treatment of lead poisoning in childhood." *Pediatrics* **44**, 291–298.

Abramowitz, D. L. 1990. "Sodium sulfide vial method for the detection of lead in paint films. Prepared for the Cambridge Housing Authority, Cambridge, Massachusetts, by Kasellan and D'Angelo Associates, Inc., Haddon Heights, New Jersey.

Alfremow, L. C. 1969. *Infrared Spectroscopy—Its Use in the Coatings Industry*, Federation of Societies for Paint Technology, Philadelphia.

Albertson, C. E. 1964. "A light tent for photomicrography." *Microscope* **14**, 253–256.

Allmand, T. R.; and Houseman, D. H. 1970. "Thin film interference—a new method for identification of nonmetallic inclusions." *Microscope* **18**, 11–23.

"Alphabetical and grouped numerical index of X-ray diffraction data." 1955. *Am. Soc. Testing Materials, Spec. Tech. Publ.* 1955,48E.

Amelinckx, S. 1964. *The Direct Observation of Dislocations.* Academic Press, London.

———, 1970. *Modern Diffraction and Imaging Techniques in Material Science.* North-Holland, Amsterdam.

Annest, J. L.; Pirkle, J. L.; Makus, D.; Neese, J. W.; Bayse, D. D.; and Kovar, G. G. 1983. "Chronological trend in blood lead levels between 1976 and 1980." Second National Health and Nutrition Examination survey. *N. Engl. J. Med.* **308**, 1373–77.

ASTM (American Society for Testing and Materials) 1989. *Annual Book of ASTM Standards*, Vol. 6.01, "Test method for low concentrations of lead, cadmium, and cobalt in paint by atomic absorption spectroscopy." Philadelphia. D 335.

ASTM–Wynadotte Index. "Molecular formula list of compounds, names, and references to published infrared spectra. *Am. Soc. Testing Materials, Spec. Tech. Publ.* **131** (1962), **131-A** (1963).

ATSDR (Agency for Toxic Substances and Disease Registry). 1988. "The nature and extent of lead poisoning in children in the United States: a report to congress." U. S. Public Health Service.

Babb, M. 1991. "Federal and State Regulations on Hazardous Lead Paint Waste." In *Lead Paint Removal: Meeting the Challenge—Proceedings of the Fourth Annual Conference*, Lead Paint Removal from Industrial Structures, Pp. 42–52.

Bellamy, L. J. 1958. *The Infrared Spectra of Complex Molecules.* 2d ed. Wiley, New York.

Bellinger, D.; Needleman, H. L.; Bromfield, R.; and Mintz, M. 1986. "A follow-up study of the

academic attainment and classroom behavior of children with elevated dentin lead levels."
     Biol. Trace Elem. Res. **6**, 207–23.

Bentley, Kenneth W. 1960. *The Natural Pigments*, Wiley–Interscience, New York.

Bertin, E. P. 1970. *Principles and Practice of X-Ray Spectrometric Analysis*. Plenum, New York.

Billick, I. H. 1983. "Sources of lead in the environment." In *Lead Versus Health* (M. Rutter and
     R. Jones, eds.). Wiley–Interscience, New York.

Birks, L. S.; Brooks, E. J.; and Friedman, H. 1953, *Anal. Chem.* **25**, 692.

Birks, L. S. 1959. *X-Ray Spectrochemical Analysis*, Wiley–Interscience, New York.

———, 1963. *Electron Probe Microanalysis.* Wiley–Interscience, New York.

Blackburn, A. J. 1990. "A comparison of the results of *in-situ* spot tests for the presence of lead
     in paint films using sodium sulfide with the results of laboratory measurements of lead
     concentrations in paint films using flame atomic absorption spectroscopy." Dewberry and
     Davis, contract HC-5831. U. S. Dept. of Housing and Urban Development.

Blackfan, K. D. 1917. "Lead poisoning in children with especial reference to lead as a cause of
     convulsions. *Am. J. Dis. Child* **8**, 377–80.

———, 1917. "Lead poisoning in children." *Am. J. Med Sci.* **153**, 877–87.

Boatwright, J. H. 1990. "Worldwide history of paint." In *Organic Coatings: Their Origin and
     Development* (R. A. Seymour *et al.*, eds.). Elsevier, New York. Pp. 8–10.

Bock, F. S. 1767. *Attempt at a Natural History of Prussian History*, Germany.

Boher, P.; and Garnier, C. R. 1984. *Seances Acad. Sci.* **298**, 203.

Boggess, W. R., and Wixio, B. G. (eds.), 1977. *Lead in the Environment*. National Science
     Foundation, Washington, DC.

Bosquet, J. 1962. *Bull. Soc. Fr. Mineral. Cristallogr.* **94**, 332.

Bowie, S. H. U.; and Taylor, K. 1958. "A system of ore mineral identification." *Min. Mag.* **99**,
     265–77, 337–45.

Bowin, J.-C.; Thomas, D.; and Tridot, G. 1968. *Compt. Rend.* **267C**, 532–35.

Boyde, A. 1970. "Practical problems and methods in three-dimensional analysis of scanning
     electron microscopy images." In *Published Proceedings of the Third Annual Scanning
     Electron Microscopy Symposium*. IIT Research Institute, Chicago. Pp. 107–12.

———, 1970. In *Scanning Electron Microscopy/1970* (Jahari, O., ed.). IIT Research Institute,
     Chicago. Pp. 105–12.

Bradley, D. E.; Halliday, J. S.; and Hirst, W. 1956. "Stereoscopic reflection electron microscopy."
     Proc. Phys. Soc. (London) **68**, 484–86.

Bragg, W. L. 1933. *The Crystalline State*. MacMillan, New York.

Breuer, K. H.; and Eysel, W. 1980. *JCPDS Grand-in-Aid Report*. Mineralogisch-Petrographisches
     Institut, Universität Heidelberg, Germany.

Brunekreef, B.; Noy, D.; Biersteker, K.; and Boleij, J. 1983. "Blood lead levels of Dutch
     children and their relationship to lead in the environment." *J. Air Poll. Cont. Assoc.* **33**,
     872–76.

Buerger, M. J. 1942. *X-Ray Crystallography*. Wiley, New York.

Bunn, C. W. 1961. *Chemical Crystallography*. 2d ed. Oxford University Press, New York.

Butler, W. J.; Goldstein, H. M.; Gooch, J. W.; Graef, J. W.; Landrigan, P. J.; Mielke, H.; Peters,
     E. T.; Shaheen, S. J.; Silbergeld, E. K.; Smith, J. K.; and Wixon, B. G. 1991. Depositions in
     Civil Action no. 87-2799-T. U.S. District Court for the District of Massachusetts.

Byers, R. E.; and Lord, E. E. 1943. "Late effects of lead-poisoning mental development." *Am.
     J. Dis. Child* **66**, 471–83.

Calkin, J. B. 1934. "Microscopical examination of paper." *Paper Trade J.* **99**, 267.

Cameron, E. N. 1969. *Ore Microscopy*. Wiley, New York.

Campbell paint brochure, exterior paint products. 1924. "Titan-O-Zinc, outside white."
     GLD31507.

Castaing, R. 1951. Thesis, University of Paris.

CDC (Centers for Disease Control). Revised 1991. Preventing lead poisoning in young children: A statement by the Centers for Disease Control. Atlanta.

CDC (Centers for Disease Control). 1985. Preventing lead poisoning in young children: A statement by the Centers for Disease Control. CDC report no. 99-2230.

Chaney, R. L.; Mielke, H. W.; and Sterrett, S. B. 1989. "Speciation, mobility, and bioavailability of soil lead. In *Lead in Soil: Issues and Guidelines* (Davies, B. E.; and Wixson, G. G., eds.). Science Reviews, Nowood, United Kingdom, Pp. 105–30.

Chang, C. C. 1971. "Auger electron spectros copy." *Surface Sci.* **25**, 53–59.

Chemical Week. 1954. November 13.

Clark, G. L. 1955. *Applied X-Rays.* 4th ed. McGraw-Hill, New York.

Colthup, N. B.; Daly, L. H.; and Wiberley, S. E. 1964. *Introduction to Infrared and Raman Spectroscopy.* Academic Press, New York.

Conti, L.; and Zocchi, M. 1959. *Acta Cryst.* **12**, 416.

Cooke, C. J.; and Duncumb, P. 1969. "Performance analysis of a combined electron microscope and electron probe microanalyzer, EMMA." In *Fifth International Congress on X-Ray Optics and Microanalysis* (Mollenstedt, G.; and Gaukler, K. H., eds.). Springer-Verlag, Berlin. Pp. 245–47.

Crewe, A. V. 1970. "High-resolution scanning microscopy of biological specimens." *Ber. Bunsen Ges. Phys. Chem* **74**, 1181–87.

Currie, L. A. 1988. "Detection: overview of historical, societal, and technical issues." In *Detection in Analytical Chemistry: Importance, Theory, and Practice* (Currie, L. A., ed.). ACS Symposium Series 361. American Chemical Society, Washington, DC.

Cushing, H. B. 1934. "Lead poisoning in children." *Int. Clin.* **1**, 189–95.

Dean, J. A.; and Rains, T. C., eds. *Flame Emission and Atomic Absorption Spectrometry.* Vol. 1, *Theory.* 1969. Vol. 2, *Components and Techniques.* 1971, Vol. 3, *Elements and Matrices.* 1974. Marcel Dekker, New York.

Department of Health, Education and Welfare. 1984. *NIOSH Manual of Analytical Methods.* NIOSH 7082.

De Fillippis, D. 1990. Technical report to J. W. Gooch, Radiocarbon Laboratory, Center for Applied Isotope Studies, University of Georgia, Athens, Georgia.

Dewolff, J. 1948. *Tech. Phys. Dienst.* Delft, Holland.

Diedrich Chemicals. 1988. "404 RIP–STRIP." Technical bulletin.

Dioscorides. A.D. 40–90. *Historia Naturalis.* Edited and translated by C. Sprengel. 1829–1830. Leipzig.

Drew, J. 1978. "Naval stores, the adaptable resource." *Chemistry* **51**, 17–19.

Dumond Chemicals. 1989. "The Peel Away Tile and Mastic Remover." Technical bulletin.

Dutch Boy Painter Magazine. 1939, 1940, 1943.

"Eagle–Picher history." 1943. *Engineering and Mining Journal* **144**, 11.

"Eagle–Picher: This lead and zinc house presents a case study in industrial transition." 1947. *Fortune* **1947**, 186.

"18,000 thousand gallons of paint and the story of Sherwin–Williams." 1935. *Fortune.* Pp. 75–80.

Elmore, W. C. 1938. "Ferromagnetic colloid for studying magnetic structures." *Phys. Rev.* **54**, 309–10.

EPA (U. S. Environmental Protection Agency). 1985. "Boston Lead-in-Soil Report." (Requested by Boston Department of Health and Hospitals Office of Environmental Affairs).

EPA (U. S. Environmental Protection Agency), Office of Health and Environmental Assessment, Environmental Criteria and Assessment. 1986. "Air quality criteria for lead."

EPA (U. S. Environmental Protection Agency), Office of Policy Planning and Evaluation. 1988. "Reducing lead in drinking water: a benefit analysis."

EPA (U. S. Environmental Protection Agency). 1991. "Strategy for reducing lead exposures: report to Congress." Washington, DC.

Farmer, F. J. 1968. *Paint and Varnish Technology.* Reinhold, New York. P. 31.

Feigl, F.; Anger, V.; and Oesper, R. E. 1972. *Spot Tests in Inorganic Analysis.* Elsevier, London. Pp. 285–86.

Fergurson, J. E. 1986. "Lead: petrol lead in the environment and its contribution to human blood lead levels." *Sci. Total Environ.* **50**, 1–54.

Ferraro, J. R. 1968. *Anal. Chem.* **40**, 4, 24A.

Fillippir S. D. 1990. "Technical report." Manager of Radiocarbon Laboratory, Center for Applied Isotope Studies, University of Georgia.

Gardner, H. A. 1917. *Paint Researches and Their Practical Application.* Judd and Detweiler, New York.

Gerry, E. A. 1935. *Naval Stores Handbook.* U. S. Dept. of Agr., Forest Service, Washington, DC.

Gianturco, M. 1965. In *Interpretive Spectroscopy* (Freeman, S. K., ed.). Van Nostrand Reinhold, New York. Chap. 2.

Gibson, J. L. 1904. "A plea for painted railings and painted walls of rooms as the source of lead poisoning among Queensland children." *Australasian Med. Gaz.* **23**, 149–53.

Glenn, W. 1889. "Chrome yellow considered as a poison." *Science* **13**, 347–48.

Gooch, J. W. 1991. "Evaluate and document lead paint abatement." Report to the U. S. Army Construction Engineering Research Laboratory, contract no. DACA88-90-D-006-0008, Champaign, Illinois.

————, 1980. Autoxidative crosslinking of emulsified vegetable oils and vegetable oil derived alkyds." Ph.D. Diss., University of Southern Mississippi. P. 15.

Grisafe, D. A.; and White, W. B. 1964. *Am. Mineral.* **49**, 1184–98.

Grivet, P. 1965. *Electron Optics.* Peragamon, London.

Grove, E. 1971. *Analytical Emission Spectroscopy.* Marcel Dekker, New York.

Gutfeld, R. 1992. "Lead in water of many cities is found excessive." *Wall Street Journal.* P. B7.

Haine, R.; and Cosslett, V. E. 1961. *The Electron Microscope, the Present State of the Art.* Wiley–Interscience, New York.

Hall, C. E. 1966. *Introduction to Electron Microscopy.* 2d ed. McGraw-Hill, New York.

Halliday, J. S. 1961. "Reflection electron microscopy." In *Techniques for Electron Microscopy.* Blackwell Scientific Publications, Oxford, United Kingdom. Pp. 306–24.

Hamilton, R. S.; Revitt, D. M.; and Warren, R. S. 1984. "Levels and physiochemical association of Cd, Cu, Pb, and Zn in roadside sediments." *Sci. Total Environ.* **33**, 59–74.

Hampton Paint Company. 1990. "Encapsulating Coatings." Technical bulletin.

Handbook for Wild ST4 Mirror Stereoscope. 1967. Wild, Heerbrugg. Pp. 2, 307e.

Harada, H. 1981. *J. Appl. Crystallogr.* **14**, 141.

Harrick, N. J. 1967. *Internal Reflection Spectroscopy.* Wiley, New York.

Haynes, W. 1945–1956. *History of the American Chemical Industry.* Vols. 1–6.

Heidenreich, R. D. 1964. *Fundamentals of Transmission Electron Microscopy.* Wiley–Interscience, New York.

Henke, B. L.; Newkirk, J. B.; and Mallett, G. R. eds. 1970. *Advances in X-Ray Analysis.* Vol. 13. Plenum, New York.

Herzberg, G. 1945. *Molecular Spectra and Molecular Structure.* Vols. 1, 2. Van Nostrand Reinhold, New York.

HHS (U. S. Dept. of Health and Human Services). 1991. "Strategic plan for the elimination of childhood lead poisoning." Atlanta, Georgia.

Hill, R. 1983. *J. Power Sources* **9**, 55.

Hirsh, P. B.; Howie, A.; Nicholson, B.; Pashley, D. W.; and Whelan, M. J. 1965. *Electron Microscopy of Thin Crystals*. Butterworths, London.

Hoffman, F. L. 1927. "Deaths from lead poisoning." U. S. Dept. of Labor, no. 426. Washington, DC.

Holt, L. E. 1923. "Lead poisoning in infancy." *Am. Dis. Child* **25**, 229–33.

Holt, L. E.; and Howland, J. 1923. *Diseases of Infancy and Childhood*. Appelton, New York and London.

"House that Joyce built." 1949. *Fortune*. P. 97.

Howell, P. G. T.; and Boyde, A. In "Scanning Electron Microscopy/1972 (Jahari, O.; and Corvin, I., eds.). IIT Research Institute, Chicago. Pp. 233–40.

Howie, A. 1965. In *Techniques for Electron Microscopy* (Kay, D., ed.). F. A. Davis, Philadelphia. Pp. 438–40.

Huber, J. E. *The Kline Guide to the Paint Industry*. 1978. Charles H. Kline, Fairfield, New Jersey. P. 32.

HUD (U. S. Dept. of Housing and Urban Development). 1990. "Comprehensive and workable plan for the abatement of lead-based paint in privately owned housing." A report to Congress. Washington, DC. GPO.

———, 1990. "Lead-based paint interim guidelines for hazard identification and abatement in public and Indian housing." Washington, DC. GPO.

Hurlbut, C. S., Jr. 1966. *Mineralogy*. Wiley, New York.

Hutchins, G. A. 1974. "Electron probe microanalysis." In *Characterization of Solid Surfaces* (Kane, P. F.; and Larrabee, G. B., eds.). Plenum, New York. Chap. 18.

Hydrosander. 1990. "Hydrosander Paint Removing Equipment." Technical bulletin.

Ingalls, W. R. 1908. *Lead and Zinc in the United States*. Hill, New York.

International Labor Office, series F (Industrial Hygiene), no. 11. 1927. "White lead."

Isings, J. 1961. In *Encyclopedia of Microscopy* (G. L. Clark, ed.). Reinhold, New York. P. 390.

Jacobs, M. H. 1971. "Microstructural studies with a combined electron microscope and electron probe microanalyzer (EMMA-3). In *Published Proceedings of the Twenty-Fifth Anniversary Meeting EMAG*, Inst. Phys.

Johari, O. 1971. "Total materials characterization with the scanning electron microscope." *Res./Develop.* **22**, 7.

———, 1974. "Characterization of solid surfaces." In *Scanning Microscopy* (Kane, P. F.; and Larrabee, G. B. eds.). Plenum, New York. Chap. 5.

Johari, O.; and Samuda, A. V. 1974. "Scanning electron microscopy." In *Characterization of Solid Surfaces* (Kane, P. F.; and Larrabee, G. B., eds.). Plenum, New York. Chap. 18.

Jones, S. J.; and Boyde, A. 1970. "Experimental studies on the interpretation of bone surfaces studied with SEM." In *Published Proceedings of the Third Scanning Electron Microscopy Symposium*. IIT Research Institute, Chicago. Pp. 195–200.

Kane, P. F.; and Larrabee, G. B., eds. 1974. *Characterization of Solid Surfaces*. Plenum, New York. Pp. 9–32.

Katz, G.; and Lefker, R. 1957. *Anal. Chem.* **29**, 1894.

Kay, D. 1961. *Techniques for Electron Microscopy*. Blackwell Scientific Publications, Oxford, United Kingdom.

Kirk, O. 1970. *Encyclopedia of Chemical Technology—Terpenes and Terpenoids and Rosin and Rosin Derivatives*. Wiley, New York.

Klemperer, C. 1953. *Electron Optics*. Cambridge University Press, London.

Kranzberg, M. 1967. *Technology in Western Civilization*. Vol. 1. Oxford University Press, New York.

Kwestroo, W.; and Langereis, C. 1965. *J. Inorg. Nucl. Chem.* **27**, 2533–36.

Landrigan, P. J.; and Needleman, H. L. 1991. Affidavits in Civil Action No. 87-2799-T. United States District Court for the District of Massachusetts.

"Lead-based paint hazard elimination." 1988. *Federal Register* **53**, no. 108. P. 20790.

"Lead and zinc pigments." In *Mineral Resources of the United States*, 1920–1931, and *Minerals Yearbook*, 1932–1935. Washington, DC. GPO.

Liebhafsky, H. A.; Pfeiffer, H. G.; Winslow, E. H. 1964. "X-ray methods: absorption, diffraction, and emission." In *Treatise on Analytical Chemistry*. Vol. 5, part 1 (Kolthoff, I. M.; and Elving, P. J., eds.). Wiley–Interscience, New York. Chap. 60.

Liebhafsky, H. A.; Pfeiffer, H. G.; Winslow, E. H.; and Zemana, P. D. 1960. *X-Ray Absorption and Emission in Analytical Chemistry*. Wiley, New York.

Lin Fu, J. S. 1985. "Historical perspective on health effects of lead." In *Dietary and Environmental Lead: Human Health Effects*.

Lippmann, F. 1966. *Naturwiss.* **53**, 701.

"Lithopone." 1928. In *Minerals Yearbook*. Washington, DC.

Low, M. J. D. 1969. *Anal. Chem.* **41**, 6, 97A.

———, 1970. *J. Chem. Educ.* **47**, A163, A255, A349, A415.

LTC Corporation. 1990. "Vacuum Abrasive Cleaning Equipment." Technical bulletin.

Luk, K. K.; Hodson, L. L.; Smith, D. S.; O'Rourke, J. A.; and Gutknecht, W. F. 1990. "Evaluation of lead test kits." Research Triangle Institute, contract no. 68-02-4550, Environmental Protection Agency.

MacDonald, N. C. 1971. In *Scanning Electron Microscopy* (Johari, O.; and Corvin, I., eds.). IIT Research Institute. Chicago. Pp. 89–96.

Mahaffey, K. R. 1988. "National estimates of blood lead levels: United States 1976–1980." *New Engl. J. Med.* **307**, 573–79.

Martens, C. R. 1974. *Technology of Paints, Varnishes, and Lacquers*. R. E. Krieger, Huntington, New York. Pp. 24–25, 27.

Mattiello, J. J., ed. 1941–1945. *Protective and Decorative Coatings, Paints, Varnishes, Lacquers, and Inks*. 5 Vols. Wiley, New York.

McGrath, F. E. 1969. "A metallographic preparation procedure for RDX dispersed in TNT." *Metallography* **1**, 341–47.

McKeehan, L. W.; and Elmore, W. C. 1934. "Surface magnetization in ferromagnetic crystals." *Phys. Rev.* **46**, 226–28.

McKhann, C. F. 1932. "Lead poisoning in children." *Arch. Neur. Psych.* **27**, 294–304.

McKhann, C. F.; and Vogt, E. C. 1933. "Lead poisoning in children." *JAMA* **101**131–35.

McKnight. 1991. Current studies being carried out for HUD and for the EPA.

McKnight, M. E.; Byrd, W. E.; and Roberts, W. E. 1990. "Measuring lead concentration in paint using a portable spectrum analyzer X-ray fluorescence device." NISTIR W90-650, National Institute of Standards and Technology, Gaithersburg, Maryland.

———, 1989. "Methods for measuring lead concentrations in paint films." National Institute of Standards and Technology, NISTIR 84-4209, Gaithersburg, Maryland.

McKnight, M. L. 1991. "Field measurements for measuring lead concentrations in paint films." *Proceedings of the Fourth Annual Conference—Lead Paint Removal from Industrial Structures*. Pp. 39–41.

McMurdie, H. 1986. *Powder Diffraction*. **1**, 70.

Medenback, O. 1975. *Mineral Abh.* **123**, 138.

Mellan, I. 1977. *Industrial Solvents Handbook*. Noyes Data, Park Ridge, New Jersey.

Mentzen, B. F.; and Latrach, A. 1983. *J. Appl. Crystallogr.* **16**, 430.

Metropolitan Life Insurance Company. 1930. "Chronic lead poisoning in infancy and early childhood." *Stat. Bull.* **11**(10), 4.

Meyer, H. M.; and Mitchell, A. W. 1945. *Minerals Yearbook*. Washington, DC.

Mielke, H. W.; Anderson, J. C.; Berry, K. J.; Mielke, P. W.; Chaney, R. L.; and Leach, M. 1983. "Lead concentrations in inner-city soils as a factor in the child lead problem." *Am. Journal Pub. Health* **73**, 1366–69.

Mielke, H. W.; Adams, J. L.; Reagan, P. L.; and Mielke, P. W. 1989. "Soil dust lead and childhood lead exposures as a function of city size and community traffic flows: the case for lead abatement in Minnesota." In *Lead in Soil: Issues and Guidelines* (Davies, B. E.; and Wixson, G. G., eds.). Environmental Geochemistry and Health Monograph Series 4, supplement to Vol. 9. Science Reviews, Nowood, United Kingdom. Pp. 105–30.

Minerals Resources of the United States. 1921–1928, 1931. Washington, DC: GPO.

Mineral Industry. 1903. Washington, DC: GPO. Pp. 236–37.

Minerals Yearbook. 1920–35, 1945–present. Washington, DC: GPO.

Moore, A. J. W. 1948. "A refined metallographic technique for the examination of surface contours and surface structure of metals—taper sections." *Metallurgia* **38**, 71.

Nakanishi, K. 1962. *Infrared Absorption Spectroscopy—Practical*. Holden-Day, San Francisco.

Nankivell, J. F. 1963. "The theory of electron stereo microscopy." *Optik* **20**, 171–98.

National Bureau of Standards. 1977. *Monograph*. **14**, 16.

National Bureau of Standards. 1958. *Circular 539* **8**, 32.

National Paint, Varnish, and Lacquer Association. 1951.

"National primary and secondary ambient air quality standards for lead." 1987. 40 CFR 50.12 (latest amendment 52 FR 31701)

Needleman, H. H. L.; and Gatsonis, G. 1990. "A low-level lead exposure and the IQ of children: a metaanalysis of modern studies." *JAMA* **263**, 673–78.

Needleman, H. L.; Gunnoe, C.; Leviton, A. 1979. "Deficits in psychological and classroom performance of children with elevated dentin lead levels." *N. Engl. J. Med.* **300**, 689–95.

NIOSH (National Institute of Safety and Health) 7082. 1984. *NIOSH Manual of Analytical Methods*. U. S. Department of Health, Education and Welfare, Washington, DC.

Nye, L. J. 1933. *Chronic Nephritis and Lead Poisoning*. Angus and Robertson, Australia.

Nyquist, R. P.; and Kagel, R. O. 1971. *Infrared Spectra of Inorganic Compounds*. Academic, New York.

Olby, J. K. 1966. *J. Inorg. Nucl. Cem.* **28**, 2507–12.

Olby, J. K. 1953. Technical Bulletin. Associated Lead Manufacturers Ltd. Middlesex, England.

OSHA (Occupational Safety and Health Administration). 1987. 29 CFR: Sec 1910.1025. Washington, DC: GPO.

Pannetier, G. 1964. *Bull Soc. Chim. France*. 701–05.

Patton, Temple C. 1973. *Pigment Handbook*. Wiley, New York.

Pehrson, E. W. 1930. *Mineral Resources of the United States*. part 1. Washington, DC: GPO. Pp. 113–32.

Peretti, E. A. 1957. *J. Am. Ceram. Sec.* **40**, 171.

Picklesimer, M. L. 1967. "Anodizing for controlled microstructural contrast by color." *Microscope* **15**, 472–79.

Pliny. A.D. 23–79. *Historia Naturalis*. (translated by Bostock, J.; and Riley, H. T.). Bohn's Classical Library, London, 1856–1893.

Polytechnic Press of the Polytechnic Institute of Brooklyn. 1965. *Proceedings from Symposium on System Theory, Symposium Series Volume 5*. Organized by the Microwave Research Institute of New York, Brooklyn, New York (distributed by Interscience Publishers, New York).

Pouchert, C. J. 1970. *The Aldreich Library of Infrared Spectra*. Aldrich Chemical, Milwaukee.

Pratt, H. J. 1903. *Mineral Resources*. Washington, DC: GPO. Pp. 1104–05.

Preuss, Harold P. 1974. *Pigments in Paint*. Noyes Data Corp., Park Ridge, New Jersey.

Protective and Decorative Coatings. Vol. 1, 1941. Wiley, New York.

Putley, E. H. 1970. "The pyroelectric detector." In *Semiconductors and Semimetals*. vol. 5 (Willardson, R. K.; and Beer, A. C., eds.). Academic, New York.

Robertson, J. M. 1953. *Organic Crystals and Molecules*. Cornell University Press, Ithaca, New York.

Rolfe, G. L.; and Hanley, A. 1975. *An ecosystem analysis of environmental contamination by lead*. University of Illinois at Urbana—Champaign, Institute for Environmental Studies.

Ross, J. R.; and Brown, A. 1935. "Poisonings common in children." *Canad. Publ. Health J.* **26**, 237–43.

Rothenberg, G. B. 1978. *Paint Additives: Recent Developments*. Noyes Data Corp., Park Ridge, New Jersey.

Ruddock, J. 1924. "Lead poisoning in children." *JAMA* **82**, 1682–84.

Sadtler Research Laboratories. 1963. *Catalog of Infrared Spectrograms*. Philadelphia.

Sayre, J. W. 1970. "A spot test for detecting lead in paint." *Pediatrics* **46**, 783.

Seymour, R. B. 1990. "Origin and development of polymeric coatings." In *Organic Coatings: Their Origin and Development* (Seymour, R. B.; and Mark, H. F., eds.) Elsevier, New York. P. 13.

Siebenthal, C. E.; and Stoll, A. 1920–1928. *Minerals Resources of the United States 1921*. Washington, DC: GPO. 1920. Pp. 85–96; 1922, Pp. 27–36; 1924. P. 35; 1928 (part 1), Pp. 347–62.

Siegbahn, K. 1967. *ESCA, Atomic, Molecular, and Solid State Structure Studied by Means of Electron Spectroscopy*. Almquist and Siksells, Uppsala, Sweden.

Siegbaum, K.; Nordling, C.; Johansson, G.; Hedman, J.; Heden, P. F.; Hamrin, K.; Gelius, U.; Bergmark, T.; Werme, L.; Manne, R.; and Baery, Y. 1969. *ESCA Applied to Free Molecules*. Elsevier, Amsterdam, New York.

Simonsen, J. L. *The Terpenes*. 5 vols. 1947–1952. Cambridge University Press, New York.

Skolnik, H. 1959. "The terpenes around us." *Hercules Chemist* **36**, 13–17.

————, 1968. "The literature of wood naval stores in literature of chemical technology." In *Advances in Chemistry Series*, N. 78. Am. Chem. Soc., Washington, DC.

————, 1983. "Production, processing, and utilization of naval stores." In *Handbook of Processing and Utilization in Agriculture*, vol. 2, part 2. CRC (Chemical Rubber Company) Press, Boca Raton, Florida, Pp. 467–506.

Smith, J. K. 1991. Affidavit in U.S. District Court for the District of Massachusetts, Civil Action Suite No. 87-2799-T.

Sorrell, C. 1973. *J. Am. Ceram. Soc.* **56**, 613.

Sproull, W. T. 1946. *X-Rays in Practice*. McGraw-Hill, New York.

Stark, A. D. 1982. "The relationship of environmental lead to blood lead levels in children." *Environmental Research* **27**, 372–83.

Staub, R. E.; and McCall, J. L. 1968. "Increasing the microscopic contrast of phases with similar reflectivities." *Metallography* **1**, 153–55.

Stevens, V. L.; and Lalk, R. H. 1980. "Solvent option for air quality compliance." Water-borne and Higher Solids Coatings Symposium. Sponsored by University of Southern Mississippi and Southern Society for Coatings Technology, New Orleans.

Stewart, D. D. 1895. "Lead convulsions." *Am. J. Med. Sci.* **109**, 288–306.

Stillman, M. 1960. *The Story of Alchemy and Early Chemistry*. Dover, New York.

Swanson, H. E.; and Fuyat, R. K. 1951. *Circular 539*. National Bureau of Standards.

Swanson, H. E.; and Tatge, E. 1951. *JC Fel. Reports*. National Bureau of Standards.

Swartz, W. E., Jr.; and Hercules, D. M. 1971. *Anal. Chem.* **43**, 1774.

Switzer, G.; Axelrod, J. M.; Lindberg, M. L.; and Larsen, E. S. 1948. "Tables of spacings for angle

2θ, Cu Kα, Cu Kα$_1$, Cu Kα$_2$, Fe Kα, Fe Kα$_1$, Fe Kα$_2$." Circular 29, geological survey. U. S. Dept. of the Interior, Washington, DC.

————, 1948. "Tables for conversion of X-ray diffraction angles to interplanar spacings." Publications AMS 10. Washington, DC: GPO.

Takahashi, E. 1969. *Science* **165**, 1352.

Tanguerel des Planches. 1839. *Lead Diseases.* (translated by Dana, Samuel. U. S. 1848).

Theophrastus. 372–287 B.C. *Enquiry into Planets* and *History of Stones.* (translated by Hart, A. London, 1916).

Thomas, G. 1962. *Transmission Electron Microscopy of Metals.* Wiley, New York.

Thomas, H. M.; and Blackfan, K. D. 1914. "Recurrent meningitis due to lead in a child of 5 years." *Am. J. Dis. Child.* **8**, 377–80.

Thomas, L. E. 1971. *Course Notes in Electron Microscopy.* University of Pennsylvania.

"Titanium and zinc pigments." 1928. In *Minerals Yearbook.*

Tolansky, S. 1952. "A high-resolution surface profile microscopy." *Nature* **169**, 445–46.

Turner, J. A. 1908. "Lead poisoning in childhood." *Australas Med. Congress.*

Turner, J. A. 1897. "Lead poisoning among Queensland children." *Australas Med. Gaz.* **16**, 475–79.

Tyler, P. M.; and Petar, A. V. 1929. *Mineral Resources of the United States.* Part 1. Washington, DC: GPO. Pp. 91–97.

U. S. Dept. of the Interior. 1955, 1992. "Annual survey of the American metals industry. In *Minerals Yearbook.*

United States of America before the Federal Trade Commission. Docket no. 5253. *National Lead Company et al. v. Federal Trade Commission.* 227 F. 2d 825, 7th Cir. 1953. Findings as to the facts and conclusions (issued January 12, 1953). P. 9.

US-CPSC. (U. S. Consumer Product Safety Commission). 1977. "Lead containing paint and certain consumer products bearing lead-containing paint." 16 CFR 1303.

USA Today. November 13, 1991. P. 10A.

Vacublast International. 1990. "Vacuum Abrasive Cleaning Equipment." Technical bulletin.

Vind, H. P.; and Matthews, C. W. 1976. "Field test for detecting lead-based paints." Technical note N-1445. Naval Civil Engineering Laboratory, Port Heuneme, California.

Weast, R. C. ed. 1978. *CRC Handbook of Chemistry and Physics.* 59th ed. CRC (Chemical Rubber Company) Press, Boca Raton, Florida.

Weeks, E. W. 1956. "Discovery of the elements." *J. Chem. Ed.* **1956**, 41–43.

Weismantel, G. E. 1981. *Paint Handbook.* McGraw-Hill, New York. Pp. 1-1, 1-23, 3-1, 3-23.

Wheeler, G.; and Rolfe, G. 1979. "The relationship between daily traffic volume and the distribution of lead in roadside soil and vegetation." *Environ. Poll.* **18**, 265–74.

White, J. S. 1970. *Mineral Rec.* **1**, 75.

White, W. B.; and Roy, R. 1964. *J. Amer. Ceram. Soc.* **47**, 242.

Wilkes, P. A., Jr. 1972. "A practical guide to internal reflection spectroscopy." *Am. Laboratory* **4**, 11, 42.

Willard, H. H.; Merritt, L. L., Jr.,; and Dean, J. A., eds. 1974. *Instrumental Methods of Analysis.* D. Van Nostrand, New York. Pp. 150–88.

————, 1974. *Instrumental Methods of Analysis.* 5th ed. D. Van Nostrand, New York, Pp. 350–419.

Williams, H. 1952. "Lead poisoning in young children." *Publ. Hlth. Rpts.* **67**, 230–36.

Williams, S. 1970. *Am. Mineral.* **55**, 784.

Wittry, D. B. 1964. "X-ray microanalysis by means of electron probes." In *Treatise on Analytical Chemistry.* Vol. 5, part 1 (Kolthoff, I. M.; and Elving, P. J., eds.). Wiley–Interscience, New York. Chap. 6.

Wright, C. W.; and Meyer, H. M. 1932–1933. *Minerals Yearbook*. Washington, DC: GPO.

Wyckoff, R. W. 1949. *Electron Microscopy, Technique, and Applications*. Wiley–Interscience, New York.

Young, R. D. 1971. "Surface microtopography." *Phys. Today* **24**, 42–48.

Zeitler, E. 1971. In *Scanning Electron Microscopy* (Johari, O.; and Corvin, I., eds.). IIT Research Institute, Chicago. Pp. 25–32.

Zero Company. 1985. "Vacuum Abrasive Cleaning Equipment." Technical bulletin.

Zinkel, D. F. 1975. "Chemicals from trees." *Chem. Tech.* **5**, 235–41.

Index